PARADOXES IN THE
THEORY OF RELATIVITY

PARADOXES IN THE THEORY OF RELATIVITY

By Yakov P. Terletskii

Department of Physics
Moscow State University, Moscow

With a Foreword by
Banesh Hoffman
Queens College of the City University of New York

Translated from Russian

℗ Springer Science+Business Media, LLC 1968

Yakov Petrovich Terletskii was born in 1912. In 1936, he was graduated from the Department of Physics of Moscow State University, earned his doctorate in Physico-Mathematical Sciences in 1945 and, since 1947, has been affiliated in the theoretical physics division there. In addition to his work at the University, he has been head of a section of the Institute of Nuclear Physics of the Academy of Sciences of the USSR at Dubna and, since 1963, a professor at Lumumba University's Theoretical Physics Department. He was awarded the 1948 M. V. Lomonosov Prize in recognition of his work on induction accelerators, and in 1951 he received the State Prize for work on the origin of cosmic rays. Terletskii has published more than a hundred scientific papers in Soviet, French, and German journals on problems in statistical physics, the theory of accelerators, plasma physics, magnetohydrodynamics, the theory of the origin of cosmic rays, the theory of relativity, and the theory of elementary particles. In addition to the present book, Terletskii has written *The Dynamical and Statistical Laws of Nature* (1950) and *Statistical Physics* (1966).

The original Russian text, published for the Moscow Society of Naturalists of the Academy of Sciences of the USSR by Nauka Press in Moscow in 1966, has been corrected by the author for this edition.

Яков Петрович Терлецкий
Парадоксы теории относительности
PARADOKSY TEORII OTNOSITEL'NOSTI

ISBN 978-1-4899-2676-0   ISBN 978-1-4899-2674-6 (eBook)
DOI 10.1007/978-1-4899-2674-6

Library of Congress Catalog Card Number 68-19185

© Springer Science+Business Media New York 1968
Originally published by Plenum Press in 1968
Softcover reprint of the hardcover 1st edition 1968

FOREWORD

That Einstein's insight was profound goes without saying. A striking indication of its depth is the abundance of unexpected riches that others have found in his work — riches reserved for those daring to give serious attention to implications that at first sight seem unphysical.

A famous instance is that of the de Broglie waves. If, in accordance with Fermat's principle, a photon followed the path of least time, de Broglie felt that the photon should have some physical means of exploring alternative paths to determine which of them would in fact require the least time. For this and other reasons, he assumed that the photon had a nonvanishing rest mass, and, in accordance with Einstein's $E = h\nu$, he endowed the photon with a spread-out pulsation of the form $A \sin(2\pi E t/h)$ in the photon's rest frame.

According to the theory of relativity such a pulsation, everywhere simultaneous in a given frame, seemed absurd as a physical entity. Nevertheless de Broglie took it seriously, applied a Lorentz transformation in the orthodox relativistic tradition, and found that the simultaneous pulsation was transformed into a wave whose phase velocity was finite but greater than c while its group velocity was that of the particle. By thus pursuing Einsteinian concepts into thickets that others had not dared to penetrate, de Broglie laid the brilliant foundations of wave mechanics.

Another well-known instance of the richness of Einstein's legacy is to be found in the work of Dirac on the relativistic wave equation of the electron. Among other things, this beautiful blend of wave mechanics, matrix mechanics, and relativity yielded the electron spin as a purely relativistic effect. But it also gave rise to what seemed like unphysical solutions of negative energy. By taking these solutions seriously, Dirac came to the prediction of the positron and to the basic concept of antimatter. However, in the process he neatly contrived to avoid a direct confrontation of the possibility of particles of negative energy existing as individual physical entities.

In this book, Professor Terletskii does not avoid direct con-
frontations. Nor does he confine himself to the familiar so-called
paradoxes of the reciprocal contraction of lengths, the journeying
and stay-at-home twins, and the like. He goes much farther, il-
luminating, for example, the relativistic role of the speed of light
by deriving the Lorentz transformation without using Einstein's
postulate of the constancy of this speed. But what is most daring
is his exploration not merely of the idea of particles of negative
mass but also that of particles of imaginary mass.

Only those who have struggled on their own with concepts of
this sort can realize how difficult it is to avoid pitfalls. Old habits
of thought die hard, and it is not always easy to remember that a
particle with negative mass if pushed to the right will move to the
left; or that in a given reference frame a particle with imaginary
mass can have infinite speed, in which case its energy is zero.
That Professor Terletskii has been able to present his ideas with
elegant simplicity should not blind us to the difficulties he has
overcome. Many physicists must have toyed with such concepts
only to recoil from them. Since Professor Terletskii is one of the
few who have dared to take them seriously enough to publish them
in detail, this translation of his book is most welcome.

It is a book for the adventurous in spirit, and it deals boldly
with thermodynamical, epistemological, and other problems raised
by the concepts of particles with negative mass or imaginary mass.
Perhaps the most important fact about the book is that its un-
orthodoxy is deeply rooted in orthodoxy. It offers a natural extension of Einstein's ideas, and therein lies its powerful claim to
our serious consideration.

Banesh Hoffmann

Queens College of the
  City University of New York
April, 1968

PREFACE

It is well known that the concepts of space and time dictated by the theory of relativity are markedly different from our everyday ideas about these universal forms of existence of matter. From the point of view of our usual concepts, many of the consequences of the theory of relativity seem to be "paradoxes": The results concerning the change in length of measuring rods and the rate of moving clocks appear paradoxical. An unusual limitation is placed on the speed of propagation of signals. There are many strange and even paradoxical aspects in the relativistic laws of motion, for example, in the relations connecting energy, momentum, and mass.

This book is devoted to the exposition, examination, and clarification of such "paradoxes" from the point of view of relativistic ideas about space and time. With the help of this approach, it is possible to clarify the peculiarities of the theory of relativity and to establish the essence of relativistic concepts in the clearest possible manner.

Paradoxes (mainly of a philosophical nature) associated with the contradiction that exists between the name and the true content of the theory of relativity are examined first of all. It is established that the theory of relativity, understood as the modern theory of the interrelation between space and time, does not provide any grounds for positivistic conclusions.

An analysis of kinematic paradoxes (contraction of scales, time dilation, laws of combination of velocities and their four-dimensional interpretation) carried out in the book leads to a better understanding of the four-dimensional geometrical significance of the concepts of space and time in the theory of relativity. The examination of the paradoxes of relativistic mechanics shows that only a four-dimensional approach to the laws of nature allows us to obtain a correct understanding of the interrelated laws of conservation of energy, momentum, and proper mass that appear in the theory of relativity as a single united law.

Fundamentally new theoretical results are reached through an analysis of the paradoxes associated with the assertion that the velocity of light cannot be exceeded. It is established that the principle of causality, correctly interpreted as an expression or consequence of the law concerning the increase in entropy, does not provide an absolute prohibition of superlight velocities for the transfer of physical effects. Only macroscopic superlight signals, but not microscopically reversible processes are forbidden. The elucidation of this fact allows us to make a number of theoretical predictions concerning the possibility of the existence of fundamentally new physical objects and processes which we can attempt to detect experimentally.

Among such objects are particles of negative mass. An analysis of the results that could be derived from a discovery of such particles is made in the book. In particular, it is shown that extremely powerful energy sources may occur if particles of negative mass really exist in nature.

The problems examined in this book have formed the subject of an elective course of lectures given to third-year students at the Physics Department of Moscow State University during the spring semester of 1962. The course was repeated in 1964 with some modifications. In putting together an outline of the 1962 course, I was greatly assisted by a postgraduate student, Yu. P. Rybakov, and I would like to take this opportunity to thank him. This outline was published by the Physics Department of Moscow State University in 1962 and it was used in the writing of the present book.

I would also like to express my deep gratitude to Professor K. P. Stanyukovich for reading the manuscript and giving me valuable advice.

Ya. P. Terletskii

Moscow State University

CONTENTS

INTRODUCTION

Paradoxes, i.e., unexpected consequences or results of a theory which contradict previously established ideas, play an important part in the development of science. In order to resolve a paradox, we have to make use of very basic propositions of the theory and sometimes even to reexamine or improve its foundations. Thus, the resolution of theoretical paradoxes forms an internal cause leading to the evolution of the theory by assisting its logical construction, and sometimes even leading to the clarification of its limits of validity and opening up perspectives for further improvement.

Undoubtedly, facts obtained from experiments and observations are the principal factor governing the evolution of any theory. However, facts alone cannot confirm, improve, or change a theory if they do not lead to the confirmation, improvement, or reexamination of the logical structure of the theory. Therefore, the development of a theory greatly depends on the resolution of internal inconsistencies. On the other hand, such contradictions in a theory are most clearly detected when they arise in the form of a paradox. Thus, the analysis of theoretical paradoxes is not merely an aim in itself, but is a means for the clarification of the true content of a theory, the improvement of individual propositions, and the discovery of ways that can lead to its improvement.

In the present book we are interested in the paradoxes of the theory of relativity. We will not touch upon problems associated with the theory of the gravitational fields, although we will consider some paradoxes associated with transformations to a noninertial frame of reference, a procedure which is normally classed as belonging to the domain of the general theory of relativity, because the latter is usually understood to mean the theory of relativity extended to noninertial frames of reference in the presence of gravitational fields.

Many of the contradictions in the theory of relativity arise on account of the standard manner in which it is usually presented, according to the classical formulation which was given by Einstein. Since the appearance of Einstein's first paper, the theory of relativity has been enriched by a large number of new concepts. The main content of the theory has been established as the result of numerous applications. It has been found that some ideas which were considered as basic during the period of genesis of the theory were in reality only auxiliary devices useful in the construction of the theory. It was also found that the theory could be constructed on the basis of a variety of postulates. In other words, it has become clear that the postulates of Einstein cannot be identified with the content of the theory of relativity.

A deeper analysis of the content of the theory of relativity is particularly important at the present time, since we are on the threshold of a new phase of development which will be associated with the demolition of many theoretical concepts following penetration inside the elementary particles and discovery of new physical processes taking place in radiogalaxies, supernovae, and quasars.

We shall see below that an examination of the problem of the limitation imposed on signal velocities in the theory of relativity will lead us to a reexamination of the content of the so-called principle of causality and to the conclusion that particles possessing negative and even imaginary masses can exist in nature. On the other hand, if such particles do exist, then their discovery will lead to a radical reconstruction of the entire physical picture of the universe. In turn, this will lead to new discoveries extending the mastery of man over nature.

I.
THE NAME AND CONTENT OF
THE THEORY OF RELATIVITY

§ 1. THE CONTRADICTION BETWEEN THE NAME
AND THE CONTENT

Many paradoxes and contradictory ideas concerning the meaning of certain deductions of relativity theory arise on account of the contradiction existing between the name and the content of relativity theory.

The name "The Theory of Relativity" tends to suggest that the content of the theory is "relativity." On the other hand, relativity as the basis of the theory is difficult to distinguish from relativism, i.e., the doctrine of the relativity of knowledge, relativity in the sense of subjectivism. This interpretation of the physical theory appeals to positivists and is extensively peddled by idealistic philosophers. They see relativity theory as an example of a physical theory contradicting materialism. This leads to the generalization that modern physics allegedly contradicts dialectical materialism.

On the other hand, materialistic philosophers, believing that the content of relativity theory is indeed "relativity," attempt either to reject the theory of relativity completely on the grounds that it contradicts materialism, or put forward the notion of a "physical relativity" different from relativism as the basis of modern physics. In attempting to develop the latter point of view, they attempt to eliminate the observer from the theory and to replace him by measuring instruments, without realizing that only the combination of a measuring instrument with an observer acquires features that are fundamentally different from those of all other objects of the material universe.

These debates between philosophers become completely aimless if the theory of relativity is approached as a physical theory with a definite content. It becomes clear that the content of the theory of relativity is the physical theory of space and time which

3

describes the geometrical interrelationship between the two. In this connection, "relativity" plays a subsidiary part (sometimes even of an illustrative character) indistinguishable from "relativity" in classical mechanics and other branches of theoretical physics.

Thus, in order to eliminate erroneous ideas that tend to arise because of the name, we should examine the content of the theory as carefully as possible.

§2. THE ORIGIN OF THE NAME "THE THEORY OF RELATIVITY"

The name "The Theory of Relativity" arose from the name of the fundamental principle or postulate put forward by Poincaré and Einstein as the basis of all theoretical productions of the new theory of space and time.

The name "The Principle of Relativity" or "The Postulate of Relativity" arose from the rejection of the absolute motionless frame of reference associated with a stationary ether introduced earlier for the explanation of optical and electrodynamic phenomena.

The fact is that by the beginning of the twentieth century an erroneous idea of the necessity of the existence of an absolute motionless frame of reference associated with the electromagnetic ether became fixed in the minds of physicists developing the theory of optical and electromagnetic phenomena by analogy with the theory of elasticity. Hence, the concept of absolute motion relative to a frame of reference associated with the ether became current, a concept which contradicts the earlier principles of classical mechanics (the Galilean principle of relativity). The experiments of Michelson and other physicists, however, disproved this theory of a "motionless ether" and provided the groundwork for the formulation of the converse assertion, which then received the name of "The Principle of Relativity." This is the way in which the name was introduced and established by Poincaré and Einstein in their early papers.

In his basic paper, "On the Electrodynamics of Moving Bodies" [1], Einstein writes: ". . . the unsuccessful attempts to discover any motion of the earth relatively to a 'light medium' suggest that the phenomena of electrodynamics as well as of mechanics possess no properties corresponding to the idea of absolute rest. They suggest rather that, as has already been shown to the first order of small quantities, the same laws of electrodynamics and optics will be valid for all frames of reference for which the equations of mechanics hold good. We will raise this conjecture (the purport of which will hereafter be called 'The Principle of Relativity') to the status of a postulate. . ." (v. the collection: The Principle of Relativity. Lorentz, Poincaré, Einstein, Minkowski. Izd. ONTI, 1935, p. 134).

Poincaré [2] formulated "The Postulate of Relativity" in an analogous manner, independently of Einstein: "This impossibility of experimentally establishing the absolute motion of the earth appears to be a law of nature; we are naturally led to the adoption of this law, which we call the postulate of relativity and to accept it without any qualifications" (ibid., p. 51).

Thus, the name "The Principle of Relativity" arose as the rejection of the concepts of an absolute frame of reference and absolute motion relative to this system at the end of the nineteenth century in connection with attempts at a mechanical explanation of electromagnetic phenomena.

The fact that the name does not reflect the essence of the basic postulates of the theory and the whole of its content has been repeatedly noted by scientists who have developed the theory of relativity.

Thus, the prominent Soviet theoretician L. I. Mandel'shtam ([4], p. 172) said in his lectures on the theory of relativity: "The name 'The Principle of Relativity' is a very unfortunate one. What is being asserted is that events are independent of the unaccelerated motion of a closed-coordinate system. That this is called 'The Principle of Relativity,' as we shall see later, leads us into a fallacy."

The unfortunate nature of the name was also pointed out by one of the originators of the theory of relativity who uncovered its content in a four-dimensional geometric form. As early as 1908, Herman Minkowski asserted [3]: ". . . the word 'relativity-postulate' for the requirement of an invariance with the group G_c seems to me very feeble. Since the postulate comes to mean that only the four-dimensional world in space and time is given by phenomena, but that the projection in space and in time may still be undertaken with a certain degree of freedom, I prefer to call it the 'postulate of the absolute world' (or, briefly, the 'world-postulate')" (v. the collection: The Principle of Relativity, p. 192).

Thus, we see that the names "The Principle of Relativity" and "The Theory of Relativity" do not reflect the true content of the theory.

§3. THE THEORY OF RELATIVITY AS THE MODERN THEORY OF SPACE—TIME

The content of the theory of relativity as a four-dimensional physical theory of space and time was initially revealed by Herman Minkowski in 1908. Only on the basis of these ideas was Einstein able in 1916 to construct the general theory of space—time including the phenomenon of gravitation (the so-called "General Theory of Relativity"). A brief account of those fundamental propositions of the theory of relativity that characterize it as a four-dimensional theory of space and time is given in the present section.

The main distinction between the concepts of space and time in the theory of relativity and the concepts of Newtonian physics is the organic interrelationship between space and time. This interrelationship appears in the formulas of Lorentz for the transformation of the coordinates and time when we want to proceed from one coordinate system to another.

Each physical phenomenon occurs in space and time and cannot be represented in our consciousness otherwise than in space and time. Space and time are forms of the existence of matter. They are the universal forms in which matter exists. Matter cannot exist outside of space and time.

A concrete method for representing space and time is a frame of reference, i.e., a coordinate—time set of numbers x, y, z, t comprising an imaginary coordinate-grid, and a temporal sequence of all possible spatial and temporal points.

One-and-the-same space and time can be represented by various coordinate—time grids, i.e., by various frames of reference. Instead of the numbers x, y, z, t, space—time can be represented by the numbers x', y', z', t', where the latter numbers cannot be chosen in an arbitrary manner, but are related to x, y, z, t by definite transformation formulas which express the properties of space—time.

Thus, each possible representation of space and time is connected with a definite frame of reference Σ. The frame of reference may be associated with a real reference body, which can be any rigid body. The coordinates x, y, z may be associated with concrete points of the reference body or may be defined by other operations carried out with respect to the reference body. Instants of time t can also be associated with the readings of actual clocks situated at various points of the reference body. The reference body is obviously essential for the performance of concrete measurements of spatial and temporal relations.

However, one should not identify the reference system with the reference body, as is proposed by physicists who adhere to the operational school of philosophy. In representing events, physicists make use of any frame of reference, including those with which it is not possible to associate a real heavy rigid body able to play the part of a reference body. The basis for such an arbitrary choice is the complete equality of all conceivable frames of reference. Consequently, a choice of a frame of reference is only a choice of a method for representing space and time suitable for the mapping of the phenomenon under investigation.

Thus, the frame of reference is chosen by a cognizing observer for the representation of physical phenomena. It is chosen in a manner which depends on the formulation of the problem and is not specified as a necessarily-existing large heavy body present in any material process.

If two frames of reference Σ and Σ' have been chosen to represent the same space—time, then, as has been established in the theory of relativity, the coordinates in the systems Σ and Σ' are related in a manner such that the interval s_{12}, defined for two separate events as

$$-s_{12}^2 = (x_1 - x_2)^2 + (y_1 - y_2)^2 + (z_1 - z_2)^2 - c^2 (t_1 - t_2)^2, \qquad (3.1)$$

remains unchanged when we proceed from Σ to Σ', i.e.,

$$- s_{12}^2 = (x_2' - x_2')^2 + (y_1' - y_2')^2 + (z_1' - z_2')^2 - c^2 (t_1' - t_2')^2. \qquad (3.2)$$

In other words, s_{12} is an invariant of the Lorentz transformation connecting the coordinates and the time in Σ and Σ':

$$x' = \frac{x - vt}{\sqrt{1 - v^2/c^2}}, \; y' = y, \; z' = z, \; t' = \frac{t - \frac{v}{c^2} x}{\sqrt{1 - v^2/c^2}}. \qquad (3.3)$$

From (3.3), as well as from (3.1), (3.2), follows, in particular, the relativity of the simultaneity of spatially separated events, i.e., if for two events in the frame of reference Σ we have $t_2 = t_1$, $x_2 \neq x_1$, $y_2 = y_1$, $z_2 = z_1$, then in the frame of reference Σ', moving with velocity v, we have

$$t_2' - t_1' = - \frac{v}{c^2} (x_2' - x_1'), \; x_2' \neq x_1', \; y_2' = y_1', \; z_2' = z_1', \qquad (3.4)$$

i.e.,

$$t_2' \neq t_1'.$$

This relativity of simultaneity leads to the shrinking of a moving measuring rod and to changes in the running of a moving clock as we shall see in detail in Sections 10 and 11.

It is these properties of the space—time coordinates that reflect the essential features of the new concepts of space and time in which they are united into a single geometric entity, an entity with a specific four-dimensional pseudo-Euclidean geometry defined by (3.1) and (3.2), a geometry in which time is tightly linked with space and cannot be considered independently of the latter as can be seen from (3.4).

These same concepts lead us to very important consequences regarding laws of nature expressed as the property of covariance (i.e., immutability of form) of any physical process with respect to transformations involving the four-dimensional space—time coordinates.

The concept of space—time as a unified four-dimensional manifold is reflected in the requirement of covariance.

Physicists who make use of the theory of relativity see its real content in this way. The notion of "relativity" here acquires the sense of a possible multiplicity of space—time representations of events with an invariant content, i.e., laws of nature.

§4. THE STERILITY OF ATTEMPTS TO CONNECT THE THEORY OF RELATIVITY WITH PHILOSOPHICAL RELATIVISM

Largely as the result of the incorrect designation of the theory of relativity, its real content as the new theory of the inter-connection between space and time is frequently replaced by the notion of the complete relativity of all physical knowledge. The possibility of a free choice of the frame of reference and the rep-resentation of events is interpreted as the relativity of the content of physical theories, relativity in relation to an observer associ-ated with the frame of reference.

This is used as the foundation for the notion that an observed physical object is completely nonexistent without the observer, since it manifests itself through the use of a measuring device and this depends on the point of view of the observer.

This idea can be followed consistently only if we assume that the content of any physical theory is the establishment of relations between various acts of measurement. This point of view, how-ever, was never a creative one, since any theory progresses by means of hypotheses concerning the substance of the phenomena themselves and not the methods of measuring them.

In the realm of the theory of relativity, the support of rela-tivity in the subjectivistic sense hindered the development of new ideas and concepts of space and time. Thus, for example, the un-successful search for general relativity in the sense of Mach did not aid the development of the theory of the gravitational field as a theory of the four-dimensional Riemann continuum of nonuniform curvature.

II.

EINSTEIN'S POSTULATES AND
THE LORENTZ TRANSFORMATIONS

§ 5. THE ROLE OF EINSTEIN'S POSTULATES
AND THEIR APPARENT INCONSISTENCY

The Lorentz transformations reflecting the properties of space—time were originally obtained by Lorentz through an analysis of the electromagnetic field equations, and were then derived by Einstein on the basis of two postulates — the principle of relativity and the principle of the constancy of the velocity of light. Einstein formulated these postulates in the following manner:

1. The laws which govern the changes of state of a physical system are independent of the choice of two-coordinate systems moving relatively to each other with a uniform translational motion for the formulation of these laws.

2. Each ray of light in a coordinate system "at rest" moves with a fixed velocity c independently of whether this light ray is emitted by a stationary or moving body.

For the further development of the theory of space—time, these postulates meant, first of all, the rejection of the old ideas concerning space and time as manifolds that are not organically related to each other.

The principle of relativity, by itself, does not represent anything new as it is also contained in Newtonian physics constructed on the basis of classical mechanics. Its extension to all physical processes raised objections only from proponents of the mechanical theory of electromagnetic and light ether.

The principle of the constancy of the velocity of light likewise offered only a return to pre-ether concepts in the field of electromagnetic phenomena, i.e., it was not completely unacceptable from the point of view of Newtonian ideas on space and time.

11

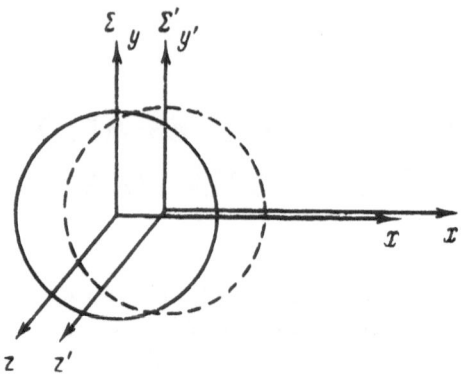

Fig. 1

However, taken together, these two principles led to contradictions with the concepts of space and time linked with Newtonian
mechanics. This contradiction is usually illustrated in the form
of the following paradox (see Fig. 1).

Let us suppose that a flash of light has occurred at the initial
time t = 0 at the origin x = y = z = 0 for a coordinate system Σ.
During the subsequent interval of time t, the front of the light wave,
on account of the law of the constancy of the velocity of light, has
become the surface of a sphere of radius R = ct whose center is at
the origin of the coordinate system Σ. However, in accordance
with Einstein's postulates, the same phenomenon can also be considered from the point of view of a coordinate system Σ' moving
uniformly and rectilinearly along the x axis, and such that, at time
t = 0, its origin and coordinate axes coincided with those of the
original system Σ. According to Einstein's postulates, in this coordinate system also, light in a time t must propagate to the surface of a sphere of radius R, although in contrast to the first sphere,
the center of the new sphere must be at the origin of system Σ' and
not Σ. The noncoincidence of these spheres, i.e., the inconsistency
between two representations of the same phenomenon, appears in
some way to be completely paradoxical and unacceptable from the
point of view of existing ideas. It seems that to resolve this paradox, we must either reject the principle of the constancy of the
velocity of light, or reject both principles.

The theory of relativity, however, suggests a completely different resolution of the paradox, namely, that events that are simultaneous in one coordinate system Σ are not simultaneous in another moving system Σ', and the converse. Then, simultaneous events consisting in the light front reaching the sphere defined by the equation

$$x^2 + y^2 + z^2 = c^2 t^2, \qquad (5.1)$$

are not simultaneous from the point of view of the system Σ' in which other events are simultaneous, namely, the light front reaching the sphere defined by the equation

$$x'^2 + y'^2 + z'^2 = c^2 t'^2. \qquad (5.2)$$

Thus, the simultaneity of spatially separated events is no longer something absolute, as is usually considered in the everyday macroscopic world, but becomes dependent on the choice of a frame of reference and the distance between the points at which the events take place. This relativity of simultaneity of spatially separated events shows that space and time are closely linked to each other, since, in proceeding from one frame of reference to another, time intervals between events become dependent on distances (a zero interval becomes finite and vice versa).

Thus, Einstein's postulates have allowed us to reach a new proposition in the physical theory of space and time, a proposition that space and time are tightly interconnected, and are not separable. This is, perhaps, the main significance of Einstein's postulates, since the derivation of the Lorentz transformations itself can be carried out in various ways with varying degrees of usage of the first and second postulates.

Some authors give a special role to the postulate of the constancy of the velocity of light, and formulate it in a manner so as to make this postulate the principal content of the theory of relativity. The main argument in support of this thesis appears to be the special role given by Einstein to light signals with the help of which the simultaneity of spatially separated events is established (i.e., determined). A light signal, always propagating at the speed of light, is thus equated to an instrument used for establishing the connection between times in different frames of reference, without which the concepts of the simultaneity of spatially separated events and time allegedly become completely meaningless.

The groundlessness of such an interpretation of the content of the theory of relativity can be easily shown if we turn to one of the possible derivations of the Lorentz transformations based only on the postulate of relativity, and instead of the postulate of the constancy of the velocity of light, we use only the assumption that the mass of a body depends on its velocity.

§6. THE MECHANICAL DETERMINATION OF SIMULTANEITY OF SPATIALLY SEPARATED EVENTS

Before we prove that the Lorentz transformations can be derived exclusively on the basis of mechanics and the principle of relativity contained in it, it is necessary to show that the simultaneity of events and the synchronization of clocks can also be established by purely mechanical experiments without the help of light signals.

One of the methods for synchronizing identical clocks situated at spatially separated points by means of light signals consists in the registration of light pulses reaching these points from a pulsed light source situated at an equal distance from the above points, for example, at the midpoint of the straight-line segment joining them. Because of the postulated constancy of the velocity of light, both pulses should simultaneously reach points equally distant from the source.

It is not difficult to see that in the given method of synchronizing clocks it is completely immaterial whether the signals propagate exactly at the velocity of light c. Here, it is only important to have them moving with the same velocity. Consequently, the light pulses can be replaced, for example, by point masses moving from the midpoint of the straight-line segment toward its ends with the same velocity u. It is only important to achieve a process of repulsion of the masses such that without further measurements we could guarantee that their velocities were the same.

The law of conservation of momentum allows us to achieve such a process, provided that the masses flying apart are the same and that no third body able to carry away a part of the momentum takes part in the repulsion process. It is obvious that many such processes of symmetric repulsion can be conceived. Hence, in principle, it is possible to realize the synchronization of spatially separated clocks by means of a purely mechanical device.

It should be emphasized that we are only speaking of possibility in principle, as in practice the use of light signals appears to be more convenient and reliable in all cases.

It may seem that the proposed mechanical-synchronization method is capable of revealing the true, nonrelativistic simultaneity of spatially separated events which is the same in all frames of reference. However, this would only be the case if the classical law of the invariance of mass were valid. In fact, experiments show that mass (defined as the ratio of momentum to the velocity) of a body depends on velocity, increasing monotonically as the latter increases. Consequently, the law of conservation of momentum becomes

$$\sum_k m_k \mathbf{u}_k = \mathbf{P} = \text{const} \tag{6.1}$$

in a "stationary" frame of reference and

$$\sum_k m'_k \mathbf{u}'_k = \mathbf{P}' = \text{const}' \tag{6.2}$$

in a "moving" frame of reference, where

$$m'_k \neq m_k, \tag{6.3}$$

since the mass appearing in (6.1) is a monotonically increasing function of the velocity, i.e., we have $m = m(\mathbf{u})$.

It is not difficult to show that, in view of (6.1)-(6.3), the simultaneous arrival at equally distant points of two bodies of equal masses, symmetrically repelled from one another in a "stationary" frame of reference, will not be simultaneous from the point of view of a moving frame of reference.

Let us suppose that two balls of equal masses are repelled by each other at time $t = 0$ according to a clock situated at the initial point x_0 and move with velocities $\neg u$ and $+u$ toward the points x_1 and x_2 situated at equal distances l from the point x_0 (see Fig. 2a). At time $t = l/u$, both balls will reach the points x_1 and x_2, at which two barriers I and II are placed (see Fig. 2b). Thus, in the

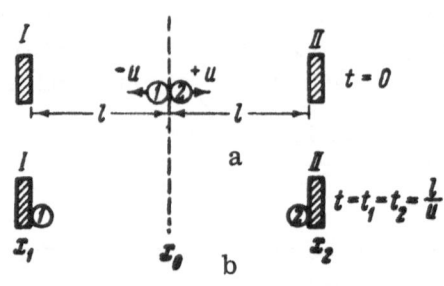

Fig. 2

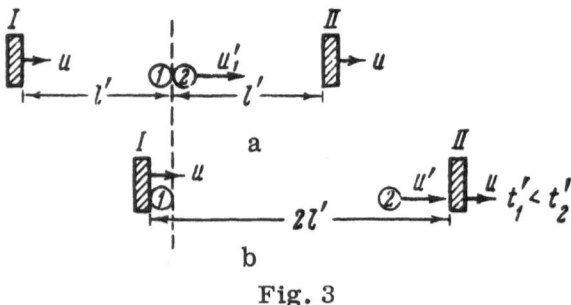

Fig. 3

chosen frame of reference the balls will simultaneously reach the barriers placed at equal distances from the initial point.

Let us now examine the same process from the point of view of another frame of reference Σ' moving to the left with velocity $\neg u$ (see Fig. 3). Relative to this frame of reference, the barriers I and II move with velocity u to the right. Up to the moment of repulsion, both balls also move to the right with velocity u and have a total momentum of $2m(u)u$. At time $t' = 0$ according to a clock situated in the frame of reference Σ' and passing at this time the point x_0, the balls are propelled away from x_0. The left ball now acquires a velocity $\neg u$, i.e., it stops moving, while the right ball acquires a velocity u', which can be found from the law of conservation of total momentum. According to this law, we have

$$P' = 2m(u)\, u = m(u')\, u',$$ (6.4)

which yields

$$u' = 2u \frac{m(u)}{m(u')},$$

(6.5)

so that velocity u' is not equal to 2u, as would be the case in classical mechanics with invariant mass.

Let the distance between barriers I and II in system Σ' be equal to $2l'$; then according to the clock in this system the left ball will reach barrier I (more correctly, barrier I will reach ball I, which in this case is stationary) at time

$$t_1' = l'/u.$$

(6.6)

The right-hand ball will reach barrier II after traversing a distance of $2l'$ at time

$$t_2' = \frac{2l'}{u'} = \frac{2l'}{2u} \cdot \frac{m(u')}{m(u)} = t_1' \frac{m(u')}{m(u)}.$$

(6.7)

Since, as the result of a monotonic increase of mass with velocity, we have

$$m(u') > m(u) \quad \text{and} \quad u' > u,$$

(6.8)

then according to (6.7) we see that

$$t_2' > t_1'.$$

(6.9)

Consequently, the principle of relativity in mechanics can be reconciled with the fact of variation of mass with velocity only if we accept the relativity of simultaneity of spatially separated events, i.e., we accept that spatially separated events that are simultaneous in one frame of reference $(t_2 = t_1)$ are not simultaneous in another moving frame of reference $(t_2' > t_1')$.

§7. THE DERIVATION OF THE LORENTZ TRANSFORMATIONS WITHOUT THE USE OF THE POSTULATE OF THE CONSTANCY OF THE VELOCITY OF LIGHT

Let us derive the Lorentz transformations solely on the basis of "natural" assumptions concerning the properties of space and time contained in classical (pre-ether) physics based on general concepts associated with classical mechanics. Let us adopt the following requirements as axioms:

1. The isotropy of space, i.e., all spatial directions are equivalent.

2. The uniformity of space and time, i.e., the independence of the properties of space and time of the choice of the initial points of measurement (the origin of the coordinates and the origin of t for time measurements).

3. The relativity principle, i.e., the complete equality of all inertial frames of reference.

The different frames of reference serve only as different representations of the one space and time as universal forms of existence of matter. Each of these representations has the same properties. Consequently, we cannot choose the relation between the coordinates and time in a "stationary" frame (x, y, z, t) and the coordinates and time in another "moving" frame (x', y', z', t'), i.e., the formulas of the coordinate and time transformations, in an arbitrary manner. Let us find the restrictions which the "natural" requirements impose on the form of the transformation functions

$$
\begin{aligned}
x' &= f_1\,(x,\,y,\,z,\,t), \\
y' &= f_2\,(x,\,y,\,z,\,t), \\
z' &= f_5\,(x,\,y,\,z,\,t), \\
t' &= f_4\,(x,\,y,\,z,\,t).
\end{aligned}
\tag{7.1}
$$

I. Because of the uniformity of space and time the transformations must be linear.

Indeed, if the derivatives of the functions f_1, f_2, f_3, f_4 with respect to x, y, z, t were not constants, but were functions of x, y, z, t, then the differences $x_2' - x_1'$, $y_2' - y_1'$, $z_2' - z_1'$, $t_2' - t_1'$, expressing the projections of the distance between points 1 and 2 in the "moving" frame of reference, would depend not only on the corresponding projections $x_2 - x_1$, $y_2 - y_1$, $z_2 - z_1$, $t_2 - t_1$ in the "stationary" frame, but also on the values of the coordinates x, y, z, t them-

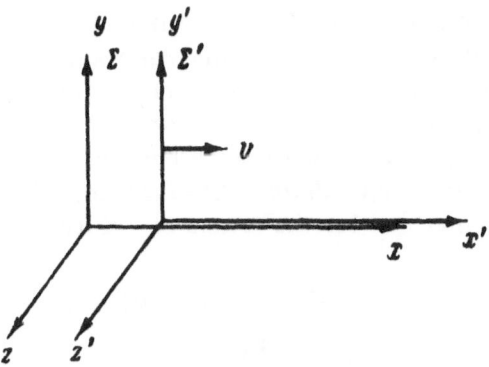

Fig. 4

selves, which would contradict the requirement that the properties of space are independent of the choice of origin of coordinates and time.

If we assume that the projections of distances of the form

$$\xi' = x_2' - x_1' = f_1(x_2, \ldots) - f_1(x_1, \ldots) \tag{7.2}$$

depend only on the projections of distances in the "stationary" frame of reference, i.e., on

$$\xi = x_2 - x_1, \tag{7.3}$$

but are independent of x_1, then

$$\partial\xi'/\partial x_1 = 0 \text{ with } \xi = \text{const}, \tag{7.4}$$

i.e.,

$$\frac{\partial f_1(x_1 + \xi, \ldots)}{\partial x_1} - \frac{\partial f_1(x_1, \ldots)}{\partial x_1} = 0, \tag{7.5}$$

or

$$\partial f_1/\partial x_1 = \text{const}. \tag{7.6}$$

We can show in an analogous manner that the derivatives of f_1 with respect to all other coordinates y_1, z_1, t_1 are also constants, and, consequently, that in general all derivatives of f_1, f_2, f_3, f_4 with respect to x, y, z, t are constants. Hence, the transformations (7.1) are linear.

II. Let us choose the "moving" frame of reference Σ' such that at the initial time $t = 0$ the point representing its origin, i.e., $x' = y' = z' = 0$, coincides with the point representing the origin of the "stationary" frame of reference, i.e., $x = y = z = 0$ (Fig. 4), while the velocity of Σ' is directed along the x axis. If we also take account of the requirement of the isotropy of space, then the linear transformations for the frame of reference Σ' can be written as

$$x' = k\,(v)(x - vt),$$
$$y' = \lambda\,(v)y,$$
$$z' = \lambda\,(v)z, \qquad\qquad (7.7)$$
$$t' = \mu\,(v)t + a\,(v)x.$$

Here there are no terms containing y and z in the expressions for x' and t' in view of the isotropy of space and the existence of a single distinguished direction along the x axis in accordance with the formulation of the problem. On the same grounds, terms proportional to z and y are absent in the expressions for y' and z', respectively, while the coefficients λ are the same for y' and z'. There are no terms containing x and t in the expressions for y' and z' because the x' axis (i.e., the line y' = z' = 0) always coincides with the x axis (i.e., the line y = z = 0). The latter would be impossible if y' and z' depended on x and t.

III. Isotropy also implies the symmetry of space. In turn, because of symmetry, nothing should change in the transformation formulas if the signs of v and x are changed, i.e., if we simultaneously change the direction of the x axis and the direction of motion of the frame of reference Σ'. Consequently, we have

$$-x' = k\,(-v)(-x + vt),$$
$$y' = \lambda\,(-v)y,$$
$$z' = \lambda\,(-v)z, \qquad\qquad (7.8)$$
$$t' = \mu\,(-v)t - a\,(-v)x.$$

Comparing (7.7) with (7.8), we obtain

$$k\,(-v) = k\,(v), \quad a\,(-v) = -a\,(v),$$
$$\mu\,(-v) = \mu\,(v), \quad \lambda\,(-v) = \lambda\,(v). \qquad\qquad (7.9)$$

Instead of the function $\alpha(v)$, it is convenient to introduce another function $\eta(v)$ such that α could be expressed in terms of η and μ as

$$\alpha(v) = -\frac{v}{\eta(v)}\mu(v). \tag{7.10}$$

According to (7.10), $\eta(v)$ is a symmetric function, i.e., $\eta(-v) = \eta(v)$.

With the help of (7.10) we can rewrite transformations (7.8) as

$$x' = k(v)(x - vt),$$
$$y' = \lambda(v)y,$$
$$z' = \lambda(v)z,$$
$$t' = \mu(v)\left[t - \frac{v}{\eta(v)}x\right], \tag{7.11}$$

in which all of the coefficients $k(v)$, $\lambda(v)$, $\mu(v)$, and $\eta(v)$ are symmetric functions.

IV. In view of the principle of relativity, both "moving" and "stationary" frames of reference are absolutely equivalent, and, hence, the inverse transformations from system Σ' to Σ must be identically equal to the direct ones (from Σ to Σ'). The inverse transformations must only differ in the sign of the velocity v, since system Σ' moves with velocity v to the right relative to the system Σ, while system Σ moves relatively to system Σ' (if the latter is considered to be stationary) to the left with velocity $-v$. Consequently, the inverse transformations should be of the form

$$x = k(-v)[x' - (-v)t'],$$
$$y = \lambda(-v)y',$$
$$z = \lambda(-v)z',$$
$$t = \mu(-v)\left[t' - \frac{(-v)}{\eta(-v)}x'\right]. \tag{7.12}$$

Comparing (7.12) with (7.11) we obtain

$$\lambda(v)\lambda(-v) = 1. \tag{7.13}$$

However, in view of symmetry $\lambda(-v) = \lambda(v)$ and, consequently, $\lambda^2 = 1$, i.e., $\lambda = \pm 1$. It is obvious that only the plus sign is meaningful, since a minus sign would give at v = 0 a system inverted with respect to y and z. Consequently, we have

$$\lambda = 1. \tag{7.14}$$

Noting that the coefficients k, μ, and η are also symmetric functions of v, the first and last of equations (7.11) and (7.12) can be written as

$$x' = k\,(x - vt). \ldots (A), \quad x = k\,(x' + vt'). \ldots (a),$$
$$t' = \mu\left(t - \frac{v}{\eta}\,x\right). \ldots (B), \quad t = \mu\left(t' + \frac{v}{\eta}\,x'\right). \ldots (b).$$

Multiplying (A) by μ, (B) by vk, and adding, we obtain

$$\mu x' + vkt' = \mu k\left(1 - \frac{v^2}{\eta}\right)x,$$
$$x = \frac{x'}{k\,(1 - v^2/\eta)} + \frac{vt'}{\mu\,(1 - v^2/\eta)}.$$

Comparing this expression with (a), we obtain

$$k = \frac{1}{k\,(1 - v^2/\eta)}, \quad k = \frac{1}{\mu\,(1 - v^2/\eta)},$$

from which we have

$$\mu = k, \quad k^2 = \frac{1}{1 - v^2/\eta}. \tag{7.15}$$

Consequently, taking the square root of the above expression and noting that the minus sign is meaningless, as in the case of λ, we obtain

$$\mu\,(v) = k\,(v) = \frac{1}{\sqrt{1 - v^2/\eta\,(v)}}. \tag{7.16}$$

Thus, the transformations become

$$x' = k\,(v)\,(x - vt), \quad y' = y, \quad z' = z,$$
$$t' = k\,(v)\left(t - \frac{v}{\eta\,(v)}\,x\right), \tag{7.17}$$

or, in more detail,

$$x' = \frac{x - vt}{\sqrt{1 - v^2/\eta}}, \quad y' = y, \quad z' = z, \quad t' = \frac{t - vx/\eta}{\sqrt{1 - v^2/\eta}}, \tag{7.18}$$

where $\eta(v)$ is, as yet, an unknown function of v.

V. In order to determine $\eta(v)$, we turn once again to the principle of relativity. It is obvious that transformations (7.17) must be universal and applicable to any transitions between one frame of reference and another. Thus, if we go from system Σ to Σ', and then from Σ' to Σ'', the formulas connecting the coordinates and time in system Σ'' with those in system Σ must be the same as the transformations (7.17). This requirement following from the principle of relativity, together with the preceding requirements of reciprocity, symmetry, etc., means that the transformations must form a *group*.

Let us make use of the group-property requirement. Let v_1 be the velocity of system Σ' relative to system Σ, and v_2 be the velocity of system Σ'' relative to system Σ'. Then, according to (7.17), we have

$$x' = k\,(v_1)\,(x - v_1 t), \qquad x'' = k\,(v_2)\,(x' - v_2 t'),$$
$$t' = k\,(v_1)\left(t - \frac{v_1 x}{\eta\,(v_1)}\right), \qquad t'' = k\,(v_2)\left(t' - \frac{v_2 x'}{\eta\,(v_2)}\right).$$

Expressing x" and t" in terms of x and t, we obtain

$$x'' = k\,(v_2)\,k\,(v_1)\left[x - v_1 t - v_2\left(t - \frac{v_1}{\eta\,(v_1)}\,x\right)\right],$$

(7.19)

$$t'' = k\,(v_2)\,k\,(v_1)\left[t - \frac{v_1}{\eta\,(v_1)}\,x - \frac{v_2}{\eta\,(v_2)}\,(x - vt)\right].$$

According to the requirement formulated above, these transformations can also be written in the form of (7.17), i.e.,

$$x'' = k\,(v_3)(x - v_3 t),$$
$$t'' = k\,(v_3)\left(t - \frac{v_3}{\eta\,(v_3)}\,x\right).$$

(7.20)

The coefficients of x in the first of formulas (7.20) and of t in the second of formulas (7.20) are the same. Consequently, in view of the identity of (7.19) and (7.20), the coefficients of x in the first of formulas (7.19) and of t in the second of formulas (7.19) must also be the same, i.e.,

$$k\,(v_2)\,k\,(v_1)\left[1 + \frac{v_1 v_2}{\eta\,(v_1)}\right] = k\,(v_2)\,k\,(v_1)\left[1 + \frac{v_2 v_1}{\eta\,(v_2)}\right].$$

(7.21)

The last equality can only be satisfied when

$$\eta\,(v_1) = \eta\,(v_2) = \text{const.}$$

(7.22)

VI. Thus, η in transformations (7.18) is a constant with the dimensions of velocity squared. The magnitude and even the sign of this constant cannot be determined without the introduction of some new assumption based on experimental facts.

If we set $\eta = \infty$, then transformations (7.18) reduce to the well-known Galilean transformations

$$x' = x - vt, \; y' = y, \; z' = z, \; t' = t. \tag{7.23}$$

These transformations, valid in the mechanics of low velocities, cannot be used as exact transformations for bodies of any velocity when the variation of mass with velocity becomes noticeable.

Indeed, as we have already seen in Section 6, when we take into account the variation of mass with velocity, we are led to the necessity of adopting the postulate of the relativity of simultaneity of spatially separated events. The latter, however, is incompatible with the Galilean transformations (7.23). Thus, the constant η must be finite.

It is known from experiment that at high velocities the equations of the mechanics of a point have the form

$$\frac{d}{dt}(m\mathbf{u}) = \mathbf{f}, \quad \text{where} \quad m = \frac{m_0}{\sqrt{1 - u^2/c^2}}; \tag{7.24}$$

where m_0 is the proper mass coinciding with the particle mass at low velocities ($v \ll c$), and c is a constant with the dimensions of velocity, numerically equal to $3 \cdot 10^{10}$ cm/sec, i.e., equal to the velocity of light in vacuo. This experimental fact is interpreted as the dependence of mass on the velocity if the mass is defined as the ratio of momentum of a body to its velocity.

The constant c^2 appearing in (7.24) has the same dimensions as the constant η appearing in the formulas for the transformation of the coordinates and time (7.18). It is therefore natural to take

$$\eta = c^2, \tag{7.25}$$

since no other constant with the dimensions of velocity squared enters into the experimentally obtained dependence of mass on velocity. Adopting (7.25), the transformation (7.18) can be written as

$$x' = \frac{x - vt}{\sqrt{1 - v^2/c^2}}, \; y' = y, \; z' = z, \; t' = \frac{t - \frac{v}{c^2}x}{\sqrt{1 - v^2/c^2}}. \tag{7.26}$$

Poincaré called these coordinate and time transformations the "Lorentz transformations."*

In view of reciprocity, the inverse Lorentz transformations must obviously be written as

$$x = \frac{x' + vt'}{\sqrt{1 - v^2/c^2}}, \; y = y', \; z = z', \; t = \frac{t' + \frac{v}{c^2}x'}{\sqrt{1 - v^2/c^2}}. \tag{7.27}$$

The dimensional analysis used by us to choose the constant η are not completely unique, since, instead of relation (7.25), we could have equally chosen

$$\eta = -c^2. \tag{7.28}$$

It is found, however, that the equations of mechanics (7.24) agreeing with experiment can only be obtained as a consequence of the Lorentz transformations (7.26) and cannot be reconciled with the transformations obtained on the basis of assumption (7.28). Indeed, it is known that the equations of mechanics based on the Lorentz transformations are the Minkowski equations† according to which mass increases with velocity according to formula (7.24). If for the coordinate transformations we choose the following:

$$x' = \frac{x - vt}{\sqrt{1 + v^2/c^2}}, \; y' = y, \; z' = z, \; t' = \frac{t + \frac{v}{c^2}x}{\sqrt{1 + v^2/c^2}}, \tag{7.29}$$

then the corresponding Minkowski equations will yield a mass m decreasing with increasing velocity, which contradicts experiment.

---

* H.A. Lorentz derived these transformations in 1904 before Poincaré and Einstein [5] (see the collection: Lorentz, Poincaré, Einstein, Minkowski, The Principle of Relativity, ONTI, 1935, p. 21). However, they were written down in the form of (7.26) for the first time by Poincaré, who called them the Lorentz transformations.

† See Section 14 for greater detail.

Thus, without appealing to the postulate of the constancy of the velocity of light, without referring to electrodynamics, and without using the properties of light signals for the determination of simultaneity, we have derived the Lorentz transformations using only the concepts of the uniformity and isotropy of space and time, the principle of relativity, and the formulas for the dependence of mass on velocity.

Usually, following the path marked out in Einstein's first paper, one replaces the formula for the dependence of mass on velocity by the postulate of the constancy of the velocity of light in vacuo. According to this postulate, in the transformation from system Σ to system Σ' the equation

$$x^2 + y^2 + z^2 - c^2 t^2 = 0 \qquad (7.30)$$

describing the front of a light wave propagating from the origin of the coordinate system Σ should remain invariant. It is easy to see that the equation

$$x'^2 + y'^2 + z'^2 - c^2 t'^2 = 0 \qquad (7.31)$$

following a transformation of the coordinates and time according to (7.18) will not change its form, i.e., (7.31) transforms into (7.30), only if $\eta = c^2$.

However, we have used another method which does not employ the postulate of the constancy of the velocity of light in order to prove that the Lorentz transformations can be derived independently of the method of signaling used for the synchronization of clocks measuring time. Physicists could be completely ignorant of the velocity of light and the laws of electrodynamics, but they could still obtain the Lorentz transformations on the basis of an analysis of the fact that mass varies with velocity and the mechanical principle of relativity. Hence, the Lorentz transformations (7.26) express the general properties of space and time for any physical process. These transformations, as was revealed in the course of the derivation, form a continuous group called the Lorentz group. It is this fact that reveals the most general properties of space and time as shown by the theory of relativity.

§8. THE DERIVATION OF THE LORENTZ TRANSFORMATIONS ON THE BASIS OF THE POSTULATE OF THE CONSTANCY OF THE VELOCITY OF LIGHT

Another derivation of the Lorentz transformations is also possible, one which is based on the postulate of the constancy of the velocity of light and the "natural" assumptions about the uniformity and isotropy of space and time. This derivation is usually adopted when it is attempted to prove that the main content of the principle of relativity is that the velocity of propagation of signals used for synchronizing clocks is limited by the velocity of light in vacuo.

The beginning of this derivation is the same as that of the preceding one. From the "natural" conditions of isotropy and uniformity we obtain the linear transformations (7.11) with symmetric coefficients $k(v)$, $\lambda(v)$, $\mu(v)$, and $\eta(v)$. Next, the requirement that the velocity of light is the same in both frames of reference is put forward.

Assuming that the front of the light signal propagates from the coordinate origin, we have for points on the wave front in both frames of reference

$$x^2 + y^2 + z^2 - c^2 t^2 = 0, \tag{8.1}$$
$$x'^2 + y'^2 + z'^2 - c^2 t'^2 = 0. \tag{8.2}$$

Substituting expressions (7.11) into (8.2), we obtain

$$(x - vt)^2 k^2 + \lambda^2 y^2 + \lambda^2 z^2 - c^2 \mu^2 \left(t - \frac{v}{\eta} x \right)^2 = 0. \tag{8.3}$$

In order for (8.3) to be consistent with (8.1), we must require that

$$k^2 = \frac{\mu^2 c^2}{\eta}, \quad k^2 - \frac{\mu^2 c^2 v^2}{\eta^2} = \lambda^2, \quad (v^2 k^2 - \mu^2 c^2) = -\lambda^2 c^2. \tag{8.4}$$

Eliminating λ and μ from these equations, we obtain

$$1 - \frac{v^2}{\eta} = \frac{\eta}{c^2} \left(1 - \frac{v^2}{\eta} \right). \tag{8.5}$$

From this, we see that $\eta = c^2$ or $\eta = v^2$. The second solution according to the first and last equations (8.4) leads to $\lambda = 0$, which is physically meaningless. Therefore, we must set

$$\eta = c^2. \tag{8.6}$$

Making use of the first and last of equations (8.4) we obtain

$$k^2 = \mu^2 = \frac{\lambda^2}{1 - v^2/c^2}. \tag{8.7}$$

Consequently, the transformations (7.11) assume the form

$$x' = \lambda(v)\frac{x - vt}{\sqrt{1 - v^2/c^2}},$$

$$y' = \lambda(v)y, \qquad z' = \lambda(v)z,$$

$$t' = \lambda(v)\frac{t - v/c^2\, x}{\sqrt{1 - v^2/c^2}}. \tag{8.8}$$

These transformations differ from the Lorentz transformations
(7.26) by the appearance of a constant scale factor $\lambda(v)$. If
$\lambda(v) \neq 1$, then, according to (8.7), there exist distinguishable systems among the inertial frames of reference. For example, the
"stationary" system Σ is distinguished by the fact that, in a transformation from it to a "moving" system Σ', the coordinates y and
z increase by a factor of λ, while in the transformation from Σ'
to Σ, y' and z' decrease by a factor of λ. The absence of a preferred system Σ clearly means that

$$\lambda(v)\lambda(-v) = 1, \quad \text{or} \quad \lambda^2 = 1. \tag{8.9}$$

However, the requirement of the absence of a preferred system in
fact denotes the requirement of the principle of relativity. Thus,
the Lorentz transformations in their final form (i.e., with $\lambda = 1$)
are obtained only after the use of a requirement which is consistent with the principle of relativity. In this derivation, however,
the principle of relativity appears to play a secondary role, since
the Lorentz transformations are obtained almost in their final form
without the use of this principle.

The comparison of the two methods for deriving the Lorentz
transformations given above confirms our assertion that the content of the theory of relativity is not the principle of relativity or
the limit imposed by the velocity of light signals, but the existence
of the Lorentz group which yields practically all of the important
consequences of the theory of relativity.

III.
PARADOXES IN KINEMATICS

§ 9. THE REPRESENTATION OF THE LORENTZ TRANSFORMATIONS IN THE MINKOWSKI PLANE

The first remarkable consequence of the Lorentz transformations is the contraction of moving scales in the direction of motion and the slowing down of moving clocks. From the point of view of everyday ideas concerning space and time, these consequences appear to be paradoxical.

An exhaustive, but somewhat formal explanation of these kinematic phenomena can be obtained in the x, ct plane if we construct in it a coordinate grid for the "stationary" and "moving" frames of reference according to the rules of Minkowski's four-dimensional geometry.

The Lorentz transformations leave the interval s_{12} between any two events, as defined by (3.1), invariant (unchanging) as can be easily seen by the substitution of (7.26) into (3.2).

Taking the first event to occur at time t = 0 at the origin of the frame of reference Σ and introducing the symmetric notation for the coordinates and time

$$x_0 = ct, \ x_1 = x, \ x_2 = y, \ x_3 = z, \tag{9.1}$$

we can write the interval between the second and first event as

$$x_0^2 - x_1^2 - x_2^2 - x_3^2 = s^2. \tag{9.2}$$

The four-dimensional geometry defined by the invariance of the interval (9.2) differs qualitatively from the usual Euclidean geometry defined by the invariance of distances, i.e., of

$$x_1^2 + x_2^2 + x_3^2 = R^2, \tag{9.3}$$

or from the simple four-dimensional generalization of Euclidean geometry in which the invariant is

$$x_1^2 + x_2^2 + x_3^2 + x_4^2 = \rho^2. \qquad (9.4)$$

In Euclidean geometries defined by (9.3) or (9.4), the square of the "distance" is always positive and, consequently, "distances" are real quantities. However, in the four-dimensional geometry defined by the interval (9.2), which provides the analog of "distance," the square of the interval may be positive, negative, or equal to zero. Correspondingly, the interval in this pseudo-Euclidean geometry can be a real or an imaginary quantity. In particular cases it may be equal to zero for noncoincident events.

It sometimes seems that the qualitative difference between the four-dimensional Euclidean geometry and four-dimensional pseudo-Euclidean geometry will disappear if, making use of Minkowski's suggestion, we assume that time is proportional to an imaginary fourth coordinate, i.e., if we set

$$x_4 = ix_0 = ict. \qquad (9.5)$$

In this case the square of the interval will become

$$x_1^2 + x_2^2 + x_3^2 + x_4^2 = - s^2, \qquad (9.6)$$

i.e., it coincides with (9.4) to within sign. However, in view of the fact that x_4 is imaginary, expression (9.6) like (9.2) can be of different sign and, hence, it differs qualitatively from expression (9.4).

In view of the invariance of intervals, the qualitative difference in the connection between events does not depend on the choice of the frame of reference, and a real, or time-like, interval ($s^2 > 0$) remains real in all frames of reference, while an imaginary, or space-like, interval ($s^2 < 0$) also remains imaginary in all frames of reference.

All the above features of pseudo-Euclidean geometry can be demonstrated by means of the Minkowski plane x_1, x_0 (see Fig. 5).

Segments Oa and Ob in this plane represent the units of scale along the time axis x_0 and the space axis x_1. The curve drawn to the right from point a is a hyperbola described by the equation

$$x_0^2 - x_1^2 = + 1, \qquad (9.7)$$

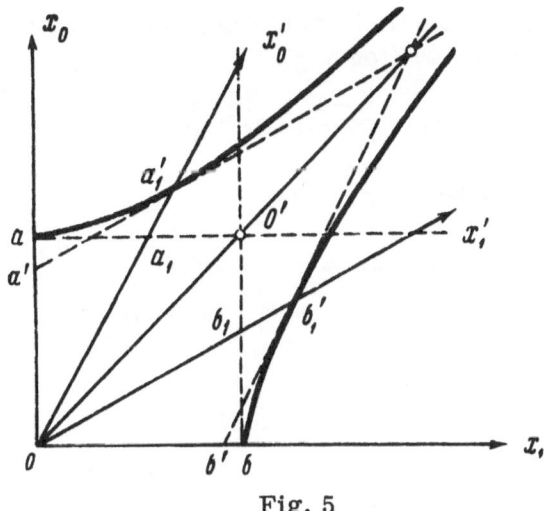

Fig. 5

while the curve drawn upward from point b is another hyperbola described by the equation

$$x_0^2 - x_1^2 = -1. \tag{9.8}$$

Thus, the coordinate origin and all points lying on the hyperbola drawn through point a are separated by unit time-like intervals. On the other hand, points lying on the hyperbola drawn through point b are separated from the coordinate origin by a space-like interval. The dashed line drawn parallel to the x_1 axis through point a represents points with coordinates $x_0 = 1$, while the line drawn parallel to the x_0 axis through point b represents points with the coordinates $x_1 = 1$.

Lines Oa_1' and Ob_1' have been drawn in the same plane to represent points with coordinates $x_1' = 0$ and $x_0' = 0$, respectively, and there are also lines drawn through a', a_1', and b', b_1' to represent points with coordinates $x_0' = 1$ and $x_1' = 1$. These lines represent the coordinate grid of system Σ'.

It can be seen from this figure that the transformation from system Σ to system Σ' corresponds in the Minkowski plane to the transformation from a rectangular to an oblique coordinate system.

This also follows directly from the Lorentz transformations (7.26), which can be written as

$$x_1' = \frac{x_1 - \beta x_0}{\sqrt{1 - \beta^2}}, \quad x_0' = \frac{x_0 - \beta x_1}{\sqrt{1 - \beta^2}}, \tag{9.9}$$

where $\beta = v/c$, or as

$$x_1' = x_1 \cosh \psi - x_0 \sinh \psi,$$
$$x_0' = -x_1 \sinh \psi + x_0 \cosh \psi, \tag{9.10}$$

where

$$\cosh \psi = \frac{1}{\sqrt{1 - \beta^2}}, \quad \sinh \psi = \frac{\beta}{\sqrt{1 - \beta^2}},$$

and, obviously,

$$\cosh^2 \psi - \sinh^2 \psi = 1.$$

Transformations (9.10) are identical with transformations from a Cartesian coordinate system to an oblique one. In these transformations, time-like vectors, i.e., vectors directed from the origin to points lying above the line OO' will remain time-like in any coordinate system, because the end points of these vectors lie on a hyperbola. Consequently, the space-like vectors will also remain space-like in all coordinate systems.

It can be seen from Fig. 5 that the "spatial" projection of the unit vector Ob_1' on the Ox_1' axis is equal to unity, while its projection on the Ox_1 axis is equal to Ob', i.e., less than unity. Consequently, a scale at rest in system Σ' will be found to be shortened when it is measured in system Σ. However, this assertion can be inverted, because the "spatial" projection of the vector Ob on the axis Ox_1' is equal to Ob_1, i.e., in system Σ' it is less than Ob_1', the unit vector.

An analogous situation is obtained in the case of "temporal" projections on the x_0 and x_0' axes. The segment Oa_1' which in system Σ' represents a process with a duration of one time unit, will in system Σ be projected as Oa', i.e., as a process with a duration less than $Oa = 1$. Consequently, the rate of a clock at rest in system Σ' will be found to be slow when measured from system Σ. It can be easily shown that this phenomenon is also reciprocal, i.e., the rate of a clock at rest in system Σ is found to be slow in system Σ'.

Let us examine these kinematic phenomena of the contraction of the four-dimensional spatial and temporal projections in greater detail, without reference to Fig. 5, on the basis only of the Lorentz transformations and the usual concepts of space, measured by means of scales, and time, measured by means of clocks.

§ 10. THE CONTRACTION OF MOVING SCALES

Although the length of a stationary scale can be measured by applying a measuring rod to it without the use of any clocks, the length of a moving scale cannot be measured directly from a stationary frame of reference without the use of clocks or signals which mark the passage of the end points of the scale being measured past the end points of the measuring rod. Therefore, by length of a moving scale we should understand the distance between its end points measured by means of a stationary measuring rod at the same time for each end. The simultaneity of the measurement of the positions of the end points is an essential condition for the experiment. It is easy to see that when this condition is not satisfied, the measured length may have any value, including a negative one or zero.

Figure 6 illustrates a scheme for the measurement of a moving scale.

Let $l_0 = x_2^! - x_1^! = l'$ be the length of a moving scale which has been measured beforehand by a direct application to a measuring rod placed in any coordinate system.* Then, if the times t_1 and t_2, at which the end points of the scale move past the points x_1 and x_2 of a stationary measuring rod, are the same (i.e., $t_1 = t_2$), then $x_2 - x_1 = l$ is by definition the length of the moving scale. Accord-

---

*Thus, we assume that the measurement of length with the help of a measuring rod is an operation which leads to the same result irrespective of further movements of the scale from one frame of reference into another. In this, of course, the scale is treated as a "rigid" body.

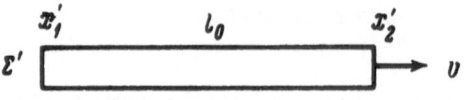

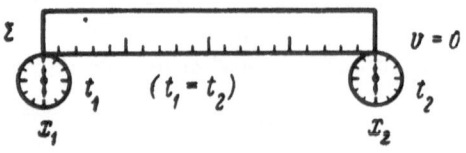

Fig. 6

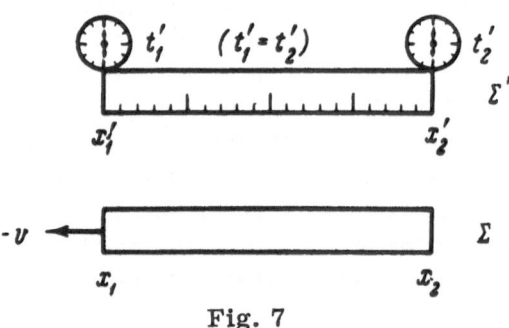

Fig. 7

ing to (7.26), we have

$$l' = x_2' - x_1' = \frac{(x_2 - x_1) - v\,(t_2 - t_1)}{\sqrt{1 - v^2/c^2}},$$ (10.1)

from which, in view of $t_1 = t_2$, we obtain

$$l = l_0\,\sqrt{1 - v^2/c^2}, \quad (l' = l_0).$$ (10.2)

The paradoxical nature of this conclusion is that in view of the principle of relativity the same formula must hold for the length of a scale situated in system Σ and measured from system Σ'. In other words, it seems that the inverse transformation should be

$$l_0 = l\,\sqrt{1 - v^2/c^2},$$ (10.3)

which is in direct contradiction to (10.2) if l_0 and l are also to be understood as measured quantities.

The contradiction, however, disappears if we take into account that relativity implies a completely symmetric change in the whole measurement scheme, i.e., the transformation from the scheme illustrated in Fig. 6 to the scheme illustrated in Fig. 7.

In this scheme, we have $t_1 \neq t_2$, i.e., the end points of the lower scale do not pass the end points of the measuring rod at the same time according to clocks situated in system Σ, but at the same time according to clocks situated in system Σ'. Then, using the formulas of the inverse Lorentz transformations, we obtain

$$x_2 - x_1 = \frac{(x_2' - x_1') + v\,(t_2' - t_1')}{\sqrt{1 - v^2/c^2}}, \qquad (10.4)$$

which, in view of $t_2' = t_1'$, yields

$$l' = l\sqrt{1 - v^2/c^2} = l_0. \qquad (10.5)$$

This formula indeed shows that there is a decrease in the length of the scale l measured from system Σ'. However, this formula no longer contradicts formula (10.2), since the quantities l and l_0 appearing in it are measured in a different way from l and l_0 appearing in (10.2).

Consequently, the contraction or expansion of measured scales depends only on the frame of reference in which we make simultaneous measurements of the end points of the scale, since events that are simultaneous in one frame of reference are not simultaneous in another.

§ 11. THE SLOWING DOWN OF MOVING CLOCKS

The slowing down of a moving clock can be observed in an experiment whose scheme is illustrated in Fig. 8. A clock, measuring time t' and moving with velocity v, passes in turn the point x_1 at time t_1 and the point x_2 at time t_2. At these times we compare the positions of the hands of the moving clock and those of stationary clocks situated next to them.*

---

*It is assumed that the simultaneity of events occurring at the same place is established directly without any signals or other

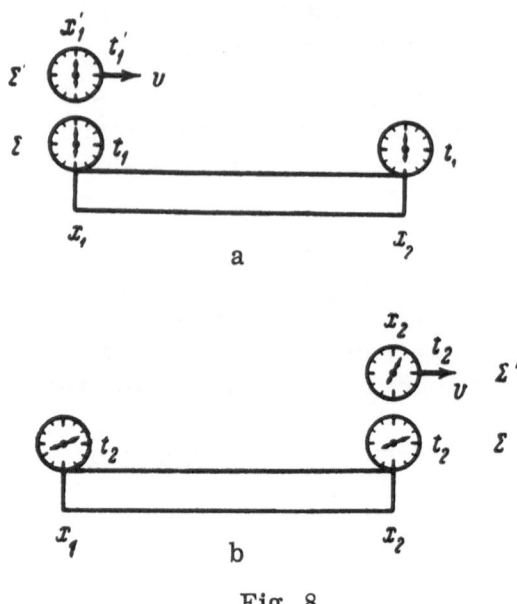

Fig. 8

Let us suppose that during the period of the motion of the clock from point x_1 to point x_2, the hands of the moving clock measure an interval of time τ_0, while the hands of clocks 1 and 2, synchronized beforehand in the stationary frame of reference Σ, measure a time interval τ. Thus, we have

$$\tau' = \tau_0 = t_2' - t_1', \qquad \tau = t_2 - t_1. \tag{11.1}$$

However, according to the inverse Lorentz transformations, we have

$$t_2 - t_1 = \frac{(t_2' - t_1') + \frac{v}{c^2}(x_2' - x_1')}{\sqrt{1 - v^2/c^2}}. \tag{11.2}$$

Substituting (11.1) into (11.2) and noting that the moving clock is always at the same point in the moving frame of reference Σ', i.e.,

---

special operations. On the other hand, the simultaneity of clocks situated at the points x_1 and x_2 in system Σ is assumed to have been established beforehand by means of light signals or the procedure examined in Section 6.

that

$$x_1' = x_2',$$ (11.3)

we obtain

$$\tau = \frac{\tau_0}{\sqrt{1 - v^2/c^2}}, \quad (\tau_0 = \tau').$$ (11.4)

This formula means that the time interval measured by a stationary clock is found to be longer than the time interval measured by a moving clock. But this also means that the moving clock lags behind the stationary one, i.e., it runs slower than the stationary clock.

Formula (11.4) is also reciprocal to the corresponding formula (10.2) for scales. However, having written the inverse formula in the form

$$\tau_0 = \frac{\tau}{\sqrt{1 - v^2/c^2}},$$ (11.5)

we must consider that $\tau_0 = t_2' - t_1'$ and $\tau = t_2 - t_1$ are now measured not in the experiment illustrated in Fig. 8, but in the experiment

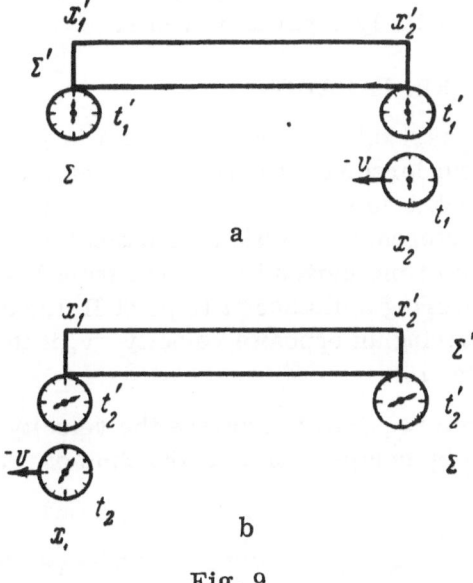

Fig. 9

illustrated in Fig. 9. In this case, according to the Lorentz transformations

$$t'_2 - t'_1 = \frac{(t_2 - t_1) - \frac{v}{c^2}(x_2 - x_1)}{\sqrt{1 - v^2/c^2}} \tag{11.6}$$

with

$$x_2 = x_1 \tag{11.7}$$

we obtain formula (11.5).

The slowing down of a moving clock is a real phenomenon*; however, it has, so to speak, a purely kinematic origin. For example, in the case of the experiment illustrated in Fig. 8, the result that clock 2 was found to be ahead of the moving clock can be explained from the point of view of the moving system Σ' by the assertion that clock 2 from the very beginning was not synchronous with clock 1 and was ahead of the latter (in view of the nonsimultaneity of spatially separated events, they are synchronous in another moving frame of reference).

Thus, both the slowing down of moving clocks and the contraction of moving scales are not paradoxical phenomena when we make ourselves familiar with the concept of the relativity of simultaneity of spatially separated events.

§ 12. THE CLOCK PARADOX

More remarkable and the cause of a large number of arguments and misunderstanding is the so-called clock paradox. Suppose that a clock A is situated at the point I in a stationary inertial frame of reference Σ, while an identical clock B, also situated at I at the initial time moves toward the point II with velocity v. Then, after covering a distance l to point II, the clock B slows down and, acquiring an opposite velocity $-v$, it then returns to point I (see Fig. 10).

If the time required to reverse the velocity of clock B is sufficiently small by comparison with the time of rectilinear and uni-

---

*It is known that rapidly moving radioactive particles, for example, μ mesons, decay with a longer half-life than stationary particles.

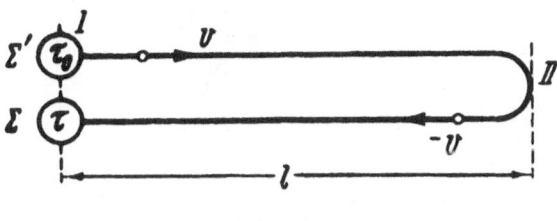

Fig. 10

form motion from point I to point II, then the time τ measured by clock A and the time τ_0 measured by clock B can be calculated according to (11.4) from the formulas

$$\tau = 2l/v, \quad \tau_0 = \tau \sqrt{1 - v^2/c^2} + \delta, \tag{12.1}$$

where δ is a possible small correction for the time during which clock B was accelerated. Consequently, clock B having returned to point I will be behind clock A by the amount

$$\Delta\tau = \tau - \tau_0 = 2\frac{l}{v}(1 - \sqrt{1 - v^2/c^2}) + \delta. \tag{12.2}$$

Since the distance l can be made as large as we please, the correction δ can in general be ignored.

The peculiarity of this kinematic consequence of the Lorentz transformation is that here the slowing down of a moving clock is an entirely real effect and not the result of the procedure chosen for measurement, as was the case in the preceding section. All processes associated with system Σ' must slow down by comparison with the processes occurring in system Σ. These will include biological processes associated with clock B. The physiological processes in a man who has traveled with system Σ' will be slowed down, as a result of which a man situated in system Σ' will have aged less when he returns to point I than a man who has remained in system Σ.

The fact that one of the clocks really lags behind the other appears to be a paradox here. For this seems to contradict the very principle of relativity, since according to the latter, either of systems Σ and Σ' can be considered to be stationary. In this case it appears that either of clocks A and B will actually slow down depending on which of the frames of reference we take to be

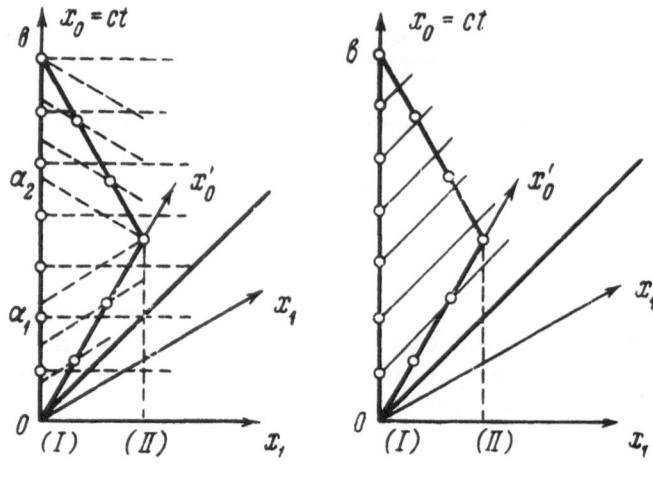

Fig. 11

the moving one. But this is clearly an absurdity, since it is actually clock B that slows down by comparison with clock A.

The error in the last argument follows from the fact that systems Σ and Σ' are not physically equivalent, because system Σ is inertial all the time, while system Σ' is noninertial during the period of time when its velocity is reversed. Consequently, the second of formulas (12.1) for system Σ' is not correct, because the rate of the moving clock can change markedly on account of the action of an inertial gravitational field at the distant point II.

However, even this completely correct explanation appears to be very remarkable. In fact, during a long period of time both systems move relatively to each other in a rectilinear and uniform manner. Therefore, from the point of view of system Σ', clock A, situated in Σ, slows down (and does not run faster) in complete agreement with formula (12.1). And only during the small interval of time when inertial forces are acting in system Σ' does clock A rapidly run ahead by an amount which is twice as great as $\Delta \tau$ calculated from formula (12.2). At this stage, the higher the acceleration suffered by system Σ', the more rapid the increase in the rate of clock A.

The significance of the above deductions can be clarified with the help of the Minkowski plane (see Fig. 11).

The segment Ob in Fig. 11a represents the stationary clock A, and the broken line Oab the moving clock B. Forces act at the point a to accelerate the system of clock B and to reverse its velocity. The points along the axis Ob subdivide the total time into equal intervals in the stationary system Σ associated with clock A.

The points along the broken line Oab mark off equal unit intervals of time measured by clock B situated in system Σ'. It can be seen from the figure that the number of unit intervals contained in the line segment Ob is greater than the number of such intervals referring to system Σ', contained in the broken line Oab. Consequently, clock B slows down by comparison with clock A.

According to the figure, the stationary clock A also lags behind clock B up to the moment represented by point a. However, after a small interval of time required for bringing clock B to rest and imparting a velocity $-v$ to it, clock B will practically show the same time A, but simultaneous to it in system Σ will now be the time α_2, i.e., almost instantaneously the time of system Σ will seemingly increase by a finite interval $\alpha_1\alpha_2$.

This "time jump," however, is not a real observable effect. Indeed, if light signals are sent at regular intervals from system Σ into system Σ', they will be received in system Σ at first at greater intervals of time and then, after the reversal of the velocity of Σ', at shorter intervals than the interval as measured in system Σ. As can be seen from Fig. 11b, there will be no discontinuity in the readings of clock A as measured in system Σ'.

Thus, the "clock paradox" is a consequence of the pseudo-Euclidean geometry of the four-dimensional space—time manifold at variance with the everyday concepts of space and time.

§13. FOUR-DIMENSIONAL AND THREE-DIMENSIONAL VELOCITY

Mass points moving rectilinearly and uniformly in the four-dimensional Minkowski space can obviously be represented by straight lines. According to Minkowski's terminology, lines rep-

resenting the motion of points are called *world lines.* Consequently, points that do not move rectilinearly and uniformly will be
represented by curved Minkowski lines. The motion at each specified point is characterized by the tangent to the world line. The
direction of the tangent can be defined as a four-dimensional vector with the components

$$dx_k/ds \ (k = 0, 1, 2, 3), \tag{13.1}$$

where

$$(ds)^2 = (dx_0)^2 - (dx_1)^2 - (dx_2)^2 - (dx_3)^2 \tag{13.2}$$

is the square of an infinitesimal interval which is a four-dimensional invariant. In accordance with Minkowski's terminology,
this vector may be called the world-tangent vector.

The components of the world-tangent vector can also be
written as

$$\frac{dx_k}{ds} = \frac{dx_k}{dx_0} \frac{1}{\sqrt{1 - (dx_1/dx_0)^2 - (dx_2/dx_0)^2 - (dx_3/dx_0)^2}} = \frac{1}{c} \frac{dx_k/dt}{\sqrt{1 - u^2/c^2}}. \tag{13.3}$$

In other words, this vector has the following four components:

$$\frac{1}{\sqrt{1 - u^2/c^2}}, \quad \frac{u_x/c}{\sqrt{1 - u^2/c^2}}, \quad \frac{u_y/c}{\sqrt{1 - u^2/c^2}}, \quad \frac{u_z/c}{\sqrt{1 - u^2/c^2}}. \tag{13.4}$$

When $u \ll c$, the zeroth component of this vector will become unity,
while the first, second, and third components form the three-
dimensional vector $\mathbf{u}/c$. Consequently, when multiplied by c, the
four-dimensional vector (13.3) can be considered to be the four-
dimensional generalization of the usual three-dimensional velocity
$\mathbf{u}$, i.e.,

$$U_k = c \frac{dx_k}{ds} = \frac{dx_k}{dt \sqrt{1 - u^2/c^2}} = \frac{dx_k}{d\tau}, \tag{13.5}$$

where

$$dt \sqrt{1 - u^2/c^2} = d\tau \tag{13.6}$$

is an element of "proper time" measured according to (11.4) by a
clock moving together with the mass point.

According to (13.2) and (13.5), the length of the four-dimensional vector U_k is obviously given by*

$$U_k U_k = U_0^2 - U_1^2 - U_2^2 - U_3^2 = c^2. \tag{13.7}$$

In four-dimensional pseudo-Euclidean geometry only the vector (13.5) can be considered as the four-dimensional velocity, i.e., a covariantly introduced physical quantity which is the generalization of the three-dimensional velocity. The conventional velocity **u** is not a covariant physical quantity since it does not comprise a quantity transforming as a four-dimensional velocity under the Lorentz transformations.

Only at low velocities ($u \ll c$), when the Lorentz transformations reduce in the limit to the Galilean transformations, does the vector **u** become a physically defined quantity covariant with respect to the Galilean transformations and rotations of coordinate axes in space.

The components of the vector U_k, like those of any other four-dimensional vector, behave in transformations from one frame of reference to another moving with velocity v in exactly the same manner as the components of dx_k,† i.e., according to the Lorentz transformations [see expressions (9.9)]

$$U_1' = \frac{U_1 - \beta U_0}{\sqrt{1 - \beta^2}}, \ U_2' = U_2, \ U_3' = U_3, \ U_0' = \frac{U_0 - \beta U_1}{\sqrt{1 - \beta^2}}, \tag{13.8}$$

where $\beta = v/c$. The "length" of the velocity vector remains constant and equal to ic.

---

* The summation convention with respect to an index repeated twice is used here and in the following; for example, the scalar product of two vectors is written as

$$A_k B_k = \sum_{k=0}^{3} \varepsilon_k A_k B_k = A_0 B_0 - A_1 B_1 - A_2 B_2 - A_3 B_3,$$

where

$$\varepsilon_0 = 1, \ \varepsilon_1 = -1, \ \varepsilon_2 = -1, \ \varepsilon_3 = -1.$$

† In the cases under consideration, there is no difference between transformations of covariant and contravariant vectors, so that we do not introduce this distinction here.

Analogously to the introduction of the four-dimensional velocity, we can introduce the four-dimensional acceleration with the components

$$\frac{d^2 x_k}{d\tau^2} = \frac{dU_k}{d\tau} .$$
(13.9)

It is easy to see that when $v \ll c$, the first three components of the four-dimensional acceleration vector coincide with those of the three-dimensional acceleration vector.

According to (13.7) we have

$$U_k \frac{dU_k}{d\tau} = 0,$$
(13.10)

i.e., the four-dimensional acceleration vector is always perpendicular to the four-dimensional velocity vector. The components of the four-dimensional acceleration vector also transform according to formulas (13.8).

For an arbitrary vector A_k, these formulas can be written as

$$A'_k = a_{k\alpha} A_\alpha,$$
(13.11)

where the matrix $\| a_{ik} \|$ is given by

$$\| a_{ik} \| = \begin{Vmatrix} \frac{1}{\sqrt{1-\beta^2}} & \frac{-\beta}{\sqrt{1-\beta^2}} & 0 & 0 \\ \frac{-\beta}{\sqrt{1-\beta^2}} & \frac{1}{\sqrt{1-\beta^2}} & 0 & 0 \\ 0 & 0 & 1 & 0 \\ 0 & 0 & 0 & 1 \end{Vmatrix}.$$
(13.12)

Transformations of the four-dimensional velocity vector and other four-dimensional vectors according to formulas (13.8), (13.12) do not lead to any paradoxical conclusions more unusual than the consequences of the Lorentz transformations which we have derived for the coordinates and time. Some new surprises only appear when we return from four-dimensional vectors to three-dimensional ones, for example, to the three-dimensional velocity vector **u**.

Any four-dimensional vector A_k can be represented as the composition of a three-dimensional part **A** with components A_1,

A_2, A_3 and a zeroth component A_0, i.e.,

$$(A_0, A) = (A_0, A_1, A_2, A_3).\qquad(13.13)$$

According to this formulation and formulas (13.4), (13.5), the four-dimensional velocity vector can be written as

$$(U_0, U) = \left(\frac{c}{\sqrt{1 - u^2/c^2}}, \frac{u}{\sqrt{1 - u^2/c^2}} \right).\qquad(13.14)$$

Consequently, the usual three-dimensional velocity vector u, measured by operations well known in classical physics, can be represented in terms of the components of the four-dimensional velocity vector as

$$u = c\frac{U}{U_0}.\qquad(13.15)$$

From this formula and the transformations (13.8) we can easily obtain formulas for the components of the vector u in a moving frame of reference Σ'. Clearly, we have

$$u'_x = c\frac{U'_1}{U'_0} = c\frac{U_1 - (v/c)U_0}{U_0 - (v/c)U_1} = \frac{u_x - v}{1 - vu_x/c^2},$$

$$u'_y = c\frac{U'_2}{U'_0} = c\frac{U_2\sqrt{1 - v^2/c^2}}{U_0 - (v/c)U_1} = \frac{u_y\sqrt{1 - v^2/c^2}}{1 - vu_x/c^2},\qquad(13.16)$$

$$u'_z = c\frac{U'_3}{U'_0} = c\frac{U_3\sqrt{1 - v^2/c^2}}{U_0 - (v/c)U_1} = \frac{u_z\sqrt{1 - v^2/c^2}}{1 - vu_x/c^2}.$$

Thus, we have obtained the so-called "formulas for the composition of velocities." It is not difficult to show that these formulas are invertible, i.e.,

$$u_x = \frac{u'_x + v}{1 + vu'_x/c^2},$$

$$u_y = \frac{u'_y\sqrt{1 - v^2/c^2}}{1 + vu'_x/c^2},$$

$$u_z = \frac{u'_z\sqrt{1 - v^2/c^2}}{1 + vu'_x/c^2}.\qquad(13.17)$$

These formulas, in particular, reveal that, by a transformation from one inertial frame of reference to another, it is impossible to obtain a velocity of a moving body that exceeds the velocity of light c, provided that the velocity of this body in the initial frame of reference is less than c.

In fact, according to formulas (13.17), the square of the three-dimensional velocity **u** transforms to

$$u'^2 = u_x'^2 + u_y'^2 + u_z'^2 = \frac{(u_x - v)^2 + (u_y^2 + u_z^2)(1 - v^2/c^2)}{(1 - vu_x/c^2)^2}. \tag{13.18}$$

Subtracting c^2 from both sides of this equation, we obtain after a simple transformation

$$u_x'^2 + u_y'^2 + u_z'^2 - c^2 = (u_x^2 + u_y^2 + u_z^2 - c^2)\frac{(1 - v^2/c^2)}{(1 - vu_x/c^2)^2}. \tag{13.19}$$

The Lorentz transformations, obviously, only have a physical meaning when v < c, since otherwise (i.e., when v > c) the primed coordinates become imaginary. Consequently, only frames of reference moving relative to the initial frame with velocities not exceeding the velocity of light have any physical meaning.* Therefore, the factor on the right-hand side of (13.19) is always positive, i.e.,

$$\frac{(1 - v^2/c^2)}{(1 - vu_x/c^2)^2} > 0, \tag{13.20}$$

which together with (13.19) means that the expression

$$u^2 - c^2 \tag{13.21}$$

does not change sign for any transformation from one frame of reference to another. Consequently, for any frame of reference we have

$$\begin{aligned} u'^2 &< c^2, \quad \text{if} \quad u^2 < c^2, \\ u'^2 &= 0, \quad \text{if} \quad u^2 = 0, \\ u'^2 &> c^2, \quad \text{if} \quad u^2 > c^2. \end{aligned} \tag{13.22}$$

In other words, sublight velocities (u < c) of any physical process remain sublight velocities in all inertial frames of reference,

---

*It should be emphasized that here we are talking about f r a m e s of r e f e r e n c e and not physical processes for which velocities exceeding c are not ruled out.

hyperlight velocities (u > c) always remain hyperlight velocities, while processes propagating at the velocity of light in any frame of reference have a velocity equal to that of light in accordance with the postulate of the constancy of light in vacuo.

IV.

PARADOXES IN RELATIVISTIC DYNAMICS

§ 14. THE MECHANICS OF A MASS POINT

Relativistically covariant equations of the motion of a mass point are the Minkowski equations, which are the natural generalization of Newton's equations, the three-dimensional velocities and accelerations being replaced by the corresponding four-dimensional quantities. Such a replacement does not contradict the reliably checked Newtonian mechanics, since at velocities for which the latter is valid, the first three components of the four-dimensional velocity and acceleration do not differ from the components of the three-dimensional velocity and acceleration.

The Minkowski equations are of the form

$$M \frac{d^2x_k}{d\tau^2} = F_k, \qquad (14.1)$$

or

$$M \frac{dU_k}{d\tau} = F_k. \qquad (14.2)$$

Here M is the proper mass, which is a four-dimensional invariant and which represents a natural generalization of the usual Newtonian mass, and F_k are the components of four-dimensional force which represents a generalization of the three-dimensional force **f**.

Introducing the four-dimensional momentum vector

$$P_k = MU_k, \qquad (14.3)$$

we can write the Minkowski equations for

$$dP_k/d\tau = F_k. \qquad (14.4)$$

In order to reveal the physical meaning of the zeroth component of the Minkowski equations and the zeroth component of the

four-dimensional momentum, as well as to establish the connection with usual three-dimensional physical measurements, we can
also write Minkowski's equations in noncovariant three-dimensional form.

Taking into account the fact that the four-dimensional velocity is perpendicular to the four-dimensional acceleration, i.e.,
taking condition (13.10) into account, we find from (14.2) that

$$F_k U_k = 0, \tag{14.5}$$

or, in the notation of (13.14),

$$F_0 U_0 - FU = 0. \tag{14.6}$$

Next, introducing the abbreviations

$$m = \frac{M}{\sqrt{1 - u^2/c^2}} \; ; \tag{14.7}$$

$$f = \sqrt{1 - u^2/c^2} \, F, \tag{14.8}$$

we find instead of (14.2) the following equations from (14.6):

$$\frac{d}{dt}(m\mathbf{u}) = \mathbf{f}, \quad \frac{d}{dt}(mc^2) = (\mathbf{u}\mathbf{f}), \tag{14.9}$$

the first of which, with $u \ll c$, reduces to the usual three-dimensional Newtonian vector equation

$$\frac{d}{dt}(M\mathbf{u}) = M\frac{d\mathbf{u}}{dt} = \mathbf{f}, \tag{14.10}$$

while the second reduces to the energy equation

$$\frac{d}{dt}\left(\frac{Mu^2}{2}\right) = (\mathbf{u}\mathbf{f}), \tag{14.11}$$

since for $u \ll c$, we have

$$mc^2 = \frac{Mc^2}{\sqrt{1 - u^2/c^2}} = Mc^2 + Mu^2/2 + \cdots \tag{14.12}$$

Thus, it has been established that the first three components
of the four-dimensional momentum is a generalization of the three-
dimensional momentum, while the zeroth component has the signifi-

cance of generalized energy divided by the velocity of light. Consequently, we have

$$\mathbf{P} = M\mathbf{U} = m\mathbf{u} = \frac{M\mathbf{u}}{\sqrt{1 - u^2/c^2}} = \frac{\mathbf{u}E}{c^2},$$

$$P_0 = MU_0 = mc = \frac{Mc}{\sqrt{1 - u^2/c^2}} = E/c. \tag{14.13}$$

In relativistic four-dimensional mechanics, the quantity $\mathbf{P}$ is called the momentum and the quantity E the energy of a mass point. The quantity m defined by (14.7) is called the inertial mass since it enters the equations of motion (14.9) in the same way as inertial mass enters the Newtonian equations of motion. Since the quantity m depends on the velocity, it is frequently called the relativistic mass.

According to (14.13), energy can be expressed in the form

$$E = mc^2 \tag{14.14}$$

or

$$E = \frac{Mc^2}{\sqrt{1 - u^2/c^2}}. \tag{14.15}$$

This is a derivation of the famous Einstein formula which is sometimes interpreted as an expression of the equivalence of mass and energy or as the law of the "inertial nature of energy."

From the point of view of the four-dimensional geometry of space—time, only four-dimensional covariant quantities can have a physical meaning. In the mechanics of mass points, such quantities are the four-dimensional scalar M, the proper mass, and the four-dimensional vectors of velocity U_k, acceleration $dU_k/d\tau$, and momentum P_k. Energy has no autonomous physical meaning in the four-dimensional theory, in contrast to the three-dimensional Newtonian theory. In four-dimensional theory, energy E is the zeroth component of the four-dimensional momentum P_k multiplied by c. Hence, energy $E = cP_0$ has a covariant physical meaning only when considered jointly with the first three components of the momentum $\mathbf{P}$. From this four-dimensional point of view, proper mass is merely a quantity proportional to the "absolute length" of the four-dimensional momentum, since according to (14.3) and (13.7), we have

$$P_k P_k = M^2 U_k U_k = M^2 c^2 \qquad (14.16)$$

or

$$(E/c)^2 - \mathbf{P}^2 = M^2 c^2. \qquad (14.17)$$

Equation (14.17) can be considered as a definition of the covariant proper mass of a body in terms of the four-dimensional momentum vector.

Thus, from the four-dimensional geometric point of view, the momentum **P** and the energy E divided by the velocity of light c (i.e., $P_0 = E/c$) form the components of a single physical quantity — the four-dimensional momentum vector. The invariant length of this four-dimensional vector is equal to the proper mass of the particle M multiplied by the velocity of light c.

§ 15. THE MEANING OF THE ASSERTION THAT MASS AND ENERGY ARE EQUIVALENT

A deep physical meaning is commonly ascribed to Einstein's relation (14.14). It is interpreted as the equivalence of mass and energy. This interpretation of Einstein's formula appears to be unavoidable if the quantity m defined by formula (14.7) is considered as a relativistic generalization of the concept of inertial mass. In this case, formula (14.14) expresses the proportionality of the relativistic energy to the relativistic inertial mass. Since expression (14.14) is a universal one, we can replace E by m in all laws, and vice versa, and this expresses the equivalence of mass and energy to within a constant factor c^2 which, by a suitable choice of units, can be made equal to unity.

If expression (14.14) is given the above meaning, then it is necessary to regard it as a relativistically covariant one. However, from the four-dimensional point of view E is the zeroth component of the four-dimensional vector

$$E_k = c P_k. \qquad (15.1)$$

Therefore, the quantity m must be considered not as an invariant, but as the zeroth component of a four-dimensional vector

$$m_k = P_k/c. \qquad (15.2)$$

Consequently, if relation (14.14) is considered not as a definition of energy in terms of quantities appearing in Eq. (14.9), but as a new physical assertion, it also acquires meaning only when it is written as

$$E_\kappa = m_k c^2, \qquad (15.3)$$

an independent physical meaning being ascribed here to the vectors E_k and m_k.

It is obvious that expression (14.7) for the inertial mass m has a covariant meaning only when it is considered as the zeroth component of an expression for a four-dimensional quantity m_k in terms of M and U_k, i.e., as the zeroth component of the vector

$$m_k = M U_k / c. \qquad (15.4)$$

All the more, there is no covariant physical meaning in such a concept as "kinetic energy" defined by the expression

$$(m - M) c^2 = M c^2 \left(\frac{1}{\sqrt{1 - u^2 / c^2}} - 1 \right), \qquad (15.5)$$

which, with $u \ll c$, coincides with the classical expression $Mu^2/2$. The first term of (15.5) is the zeroth component of a four-dimensional vector, while the second term is a four-dimensional scalar. It is obvious that such a "hybrid" composed of a vector component and a scalar cannot be considered as a covariantly introduced physical quantity when we are dealing with velocities comparable to those of light. Only for velocities $u \ll c$ does the kinetic energy acquire the meaning of a three-dimensional scalar.

The quantities defined by expressions (14.7), (14.14), (14.15), (15.5), and Eqs. (14.9) have a definite physical meaning only in the case of transitions from four-dimensional representations to three-dimensional ones associated with a fixed frame of reference, i.e., when the principle of relativity and the four-dimensional nature of space—time are ignored. The latter procedure can be justified either in the case $u \ll c$, or as a substitute for the rigorous theory intended for the reconciliation of the four-dimensional theory with the common-sense three-dimensional concepts of conventional classical physics. Indeed, if relativistic covariance is ignored and the theory is constructed for a single frame of reference, then relativistic effects can be represented as a correction

to take into account the fact that the "inertial mass" m appearing
in the usual equations of mechanics (14.9) depends on the velocity
according to the "law" (14.7). Then, the definition of energy E as
the zeroth component of a four-dimensional momentum multiplied
by c can be presented as the "law of the inertial nature of energy,"
etc.

Many misunderstandings and paradoxes arising in the interpretation of the formulas of relativistic mechanics occur because
the so-called laws that can be justified only in a three-dimensional noncovariant formulation are interpreted from a relativistic
four-dimensional point of view.

In the four-dimensional theory there is no concept of inertial mass as a scalar varying with velocity, only the concept
of proper mass M indissolubly linked with momentum and
energy. Therefore, the "law of variation of inertial mass with
velocity" can only be included in the four-dimensional theory if a
generalization of "inertial mass" of the form (15.4) is introduced.
However, such a generalization is artificial, since the mass vector m_k (15.4), apart from a constant factor of $1/c$, does not differ
in any way from the four-dimensional momentum vector P_k.

The same situation occurs in the case of Einstein's law
$E = mc^2$. From the three-dimensional point of view this is indeed
a law, because it links two qualitatively different quantities, one
of which is a property of the motion, the other the property of the
inertia of matter. From the four-dimensional point of view, this
relation is only meaningful when it is written down in the form of
(15.3). But in this case it is lowered to the status of a definition
of the mass vector m_k in terms of the energy vector E_k, while
both of these vectors are physically defined only through the momentum vector P_k, which is the only quantity with a direct physical meaning.

Einstein's relation (14.14) can be given another covariant
meaning different from that of (15.3). This is the meaning with
which this relation is used in nuclear dynamics. However, we will
leave this problem until the next section where we discuss the laws
of conservation of energy and momentum for systems of particles.

§ 16. THE LAWS OF THE CONSERVATION OF ENERGY, MOMENTUM, AND TOTAL PROPER MASS

In nonrelativistic mechanics, the law of conservation of energy is obtained as a consequence of the equations of dynamics of a system of mass points

$$\frac{d}{dt}(m\mathbf{u}_k) = \mathbf{f}_k. \tag{16.1}$$

Conservation of momentum also follows from them under the condition that

$$\sum_k \mathbf{f}_k = 0. \tag{16.2}$$

In relativistic mechanics, in view of the impossibility of neglecting the finite time of propagation of interactions, condition (16.2), expressing the law that action and reaction are equal and opposite, is no longer satisfied. However, conservation laws for systems containing both point masses and fields which guarantee the interactions between particles can be obtained as a consequence of the general law of the conservation of energy—momentum in the differential form

$$\frac{\partial T_{k\alpha}}{\partial x_\alpha} = 0, \tag{16.3}$$

where $T_{k\alpha}$ is the four-dimensional energy—momentum tensor. This problem can only be fully discussed in connection with the field theory of elementary particles. Therefore, in the present section, we will restrict ourselves only to an analysis of the laws of conservation of momentum and energy, presenting them without derivation.

Let $(e_k/c, \mathbf{p}_k)$ be the four-dimensional momentum of the k-th particle and $(e^{(f)}/c, \mathbf{p}^{(f)})$ be the four-dimensional momentum of all fields. Then, for an isolated system of fields and particles we have a law of conservation of the total momentum $(E/c, \mathbf{P})$. Consequently, for a system consisting of N particles, the quantities

$$\mathbf{P} = \sum_{k=1}^{N} \mathbf{p}_k + \mathbf{p}^{(f)},$$

$$E = \sum_{k=1}^{N} e_k + e^{(f)} \tag{16.4}$$

remain unchanged, from which, according to (14.17), the total proper mass of the system, defined as

$$M = \frac{1}{c}\sqrt{(E/c)^2 - (\mathbf{P})^2},$$

(16.5)

also remains conserved.

Thus, in relativistic theory the three-dimensional laws of conservation of momentum, energy, and mass are combined into a single four-dimensional law of the conservation of the four-dimensional momentum, or the law of the conservation of energy—momentum.

In the center-of-mass system, where

$$\mathbf{P} = 0,$$

(16.6)

the law of conservation of total proper mass coincides with the law of conservation of energy (to within a constant factor c^2). In this system

$$M = \frac{1}{c^2} E_0 = \sum_{k=1}^{N} \frac{M_k}{\sqrt{1 - \beta_k^2}} + e^{(f)}/c^2.$$

(16.7)

For particle velocities low by comparison with that of light, the field energy $e^{(f)}$ becomes the energy of the binary interaction between particles, i.e., the potential energy U of the system. Thus, Eq. (16.7) with $\beta_k \ll 1$ transforms into the relation

$$\sum_{k=1}^{N} M_k + \frac{1}{c^2}\left\{ \sum_{k=1}^{N} \frac{M_k u_k^2}{2} + U \right\} = M = \text{const.}$$

(16.8)

This relation can also be written as

$$\Delta M = \frac{1}{c^2}\left\{ \sum_{k=1}^{N} \frac{M_k u_k^2}{2} + U \right\},$$

(16.9)

where the quantity

$$\Delta M = M - \sum_{k=1}^{N} M_k \qquad (16.10)$$

is called the "mass defect." Thus, formula (16.9) means that the classical mechanical energy of a body is equal to the mass defect multiplied by the square of the velocity of light.

In the absence of any transformations of one type of particle into another, the number of particles of each type remains constant, and in view of the conservation of M, the mass defect ΔM also remains constant, so that equality (16.9) acquires the meaning of a mechanical law of conservation of energy, i.e.,

$$\sum_{k=1}^{N} \frac{M_k u_k^2}{2} + U = \mathscr{E} = c^2 \Delta M. \qquad (16.11)$$

Here, the assertion that Δm = const coincides according to (16.10) with the classical mechanical law of mass conservation

$$\sum_{k=1}^{N} M_k = \text{const.} \qquad (16.12)$$

Thus, in the nonrelativistic case, in the absence of transformations of one type of particle into another type, the law of conservation of proper mass splits up into the classical law of conservation of energy and the law of conservation of mass.

On the other hand, in the presence of transformations of one type of particle into another type, the dissociation of the system into parts, or recombination, the quantity $\sum_{k=1}^{N} M_k$, and, consequently, ΔM also change, so that the mechanical energy of the system changes according to (16.11). Therefore, the classical-like formula (16.11) acquires the meaning of a relation which indicates how the energy of the system can change following the qualitative transformation of matter from one form into another. Consequently, the classical theory is supplemented by a new relation describing the possibility of the creation or destruction of mechanical energy in qualitative transformations of matter.

Expressions (16.9), (16.11) can also be written down for the case of high particle velocities in a relativistically covariant form.

Let $E_0 = Mc^2$ be the energy of the particle system when $\mathbf{P} = 0$, i.e., the proper energy of the system in the given state. Let us consider another possible state of the same system, a state in which its proper energy is a minimum corresponding to the given composition, which is defined by the choice of the M_k, i.e., the proper masses of the N parts of the system. If to this proper mass we also add the proper mass of the field energy $e^{(f)}$, defined by

$$M_{N+1} = M^{(f)} = e_0^{(f)} / c^2, \tag{16.13}$$

then the proper energy in this state according to (16.7) is

$$E_0^{(\min)} = \sum_{k=1}^{N+1} M_k c^2, \tag{16.14}$$

while the proper mass is

$$M^{(\min)} = \sum_{k=1}^{N+1} M_k. \tag{16.15}$$

The quantities M, $M^{(\min)}$, E_0, $E_0^{(\min)}$ are four-dimensional scalars, so that we can write the following covariant expression on the basis of (16.7):

$$\mathcal{E} = E_0 - E_0^{(\min)} = c^2 (M - M^{(\min)}). \tag{16.16}$$

This relativistically covariant relation has a deeper physical meaning than formula (16.7), which expresses the proportionality between proper energy and proper mass of the system. It gives the maximum amount of heat that can be liberated in a system which has an initial proper mass M when it is converted into a system of N particles with proper masses M_k possessing the smallest possible field interaction energy equal to $c^2 M^{(f)}$. Thus, the quantity $\mathcal{E}$ can be called the active energy of the system with the given composition.

According to (16.10), (16.11), and (16.15), formula (16.16) can also be written as

$$\mathscr{E} = c^2 \Delta M - U, \qquad (16.17)$$

where we have used the abbreviation

$$U = \rho_0^{(f)} = M^{(f)} c^2.$$

If in the final state the particles are so far apart that the field energy U can be neglected, then formula (16.17) assumes the simplified form

$$\mathscr{E} = c^2 \Delta M. \qquad (16.18)$$

This formula determines the maximum energy which can be obtained when an initial system with proper mass M decays into N particles with proper masses M_1, M_2, . . . , M_N.

If a system, consisting of N' particles with proper masses M_1', M_2', . . . , $M_{N'}'$, transforms into a system with N particles of proper masses M_1, M_2, . . . , M_N, then the increment in the active energy of such a system according to (16.16) will be

$$\Delta\mathscr{E} = -c^2 \Delta M^{(min)}. \qquad (16.19)$$

In this formula, it is clear that

$$\Delta M^{(min)} = \sum_{k=1}^{N'+1} M_k' - \sum_{k=1}^{N+1} M_k. \qquad (16.20)$$

If we are dealing with collections of free particles in both the initial and final states, i.e., if we can neglect their binding energies, then the field masses $M_{N'+1}'$ and M_{N+1} can be neglected, and instead of (16.20) we have

$$\Delta M^{(min)} = \sum_{k=1}^{N'} M_k' - \sum_{k=1}^{N} M_k. \qquad (16.21)$$

Formula (16.19) describes the active energy liberated in nuclear transformations or in transformations of elementary particles, i.e., in the transformation of matter from one form into another.

§ 17. DOES MASS CHANGE INTO ENERGY?

Formula (16.16) has a different physical meaning from formula (14.14), but it is also called Einstein's relation. In the form of (16.17), Einstein's relation or law is used in nuclear physics. On the basis of this relation, it is commonly asserted that the mass of a body is a measure of the energy contained in it. Frequently, with reference to formula (16.19), it is said that mass is converted into energy.

The assertion that mass can be converted into energy is met by particularly vigorous objections on the part of materialist philosophers and, conversely, it is used by idealist philosophers as a refutation of materialism. The fact is that energy is usually taken to mean the quantitative measure of motion in its transformation from one form to another. On the other hand, mass is adopted as a quantitative measure of matter.

The origins of these ideas date back to the end of the last century, and they have been used by Engels in his "Dialectic of Nature." From this point of view, the conversion of mass into energy means the destruction of matter and the generation of motion, i.e., the conversion of matter into motion. This assertion contradicts dialectical materialism, which is based on the indestructibility of matter and its motion, but is accepted without objection in idealistic philosophy. It can be used for the refutation of the basic thesis of materialism that matter exists as an objective reality, exists apart from our consciousness. In order to resolve these contradictions, let us examine whether it is possible to speak of the conversion of mass into energy from the point of view of physics.

If on the basis of the four-dimensional concept of space and time we consider energy as a quantity proportional to the zeroth component of the four-dimensional momentum, while by proper mass we mean the absolute magnitude of the four-dimensional momentum (its "length"), then the assertion of the conversion of mass into energy becomes completely meaningless because energy and mass are only different projections of one-and-the-same physical quantity, and both these quantities are conserved in all transformations. In the case of an isolated system, the total energy and the total proper mass of the system remain unchanged in any physical transformation. Consequently, in a consistent relativistic theory there is no conversion of mass into energy and formulas (14.14), (14.17), (16.5), (16.7) merely express the organic interrelationship between the concepts of mass, energy, and momentum.

In four-dimensional theory, the laws of conservations of energy−momentum and proper mass (16.4), (16.5) represent a single law which expresses both the conservation of matter in its transformations from one form into another and the indestructibility of motion.

Thus, the relativistic formulas connecting energy, momentum, and proper mass do not imply any possibility of the destruction of matter or momentum. They only lead to the necessity of combining the general law of conservation and transformation of matter and the law of conservation of momentum into a single law of nature from which follows not only the indestructibility of matter, but also the indestructibility of momentum. This new proposition, obviously, does not in any way contradict dialectical materialism, although it enriches and supplements it.

On the other hand, if we move away from the rigorous relativistic concepts of energy and proper mass and take energy to be that part of the quantity $E = cP_0$ which can be converted into heat, then formulas (16.9), (16.16) can be interpreted to mean the conversion of proper mass as a new form of energy into energy understood in the classical sense. In fact, according to (16.9), (16.16) this "classical energy" is not conserved in transformations taking place with a change in the sum of the total masses of parts of the system. According to (16.9) it increases by an amount proportional to the decrease in the sum of the proper masses of the subsystems. Thus, from this classical point of view, classical energy is not conserved when transformations of matter involve a change in the sum of the proper masses of the component parts. However, in the same way as kinetic energy is not conserved in classical mechanics, but the sum of the kinetic and potential energies is conserved, we can consider that, in the case of transformations of matter, what is conserved is classical energy plus a hidden form of energy equal to the difference between the sums of the proper masses of the system components before and after the transformation multiplied by c^2. Consequently, if $c^2 \Delta M^{(min)}$ is considered as a hidden energy, then a transformation of matter does not involve the conversion of mass into energy, but the conversion of a hidden form of energy into an active classical energy. With this interpretation, however, formula (16.19) is more correctly written as

$$\Delta \mathscr{E}_{\text{cl}} = \Delta \mathscr{E} + c^2 \Delta M^{(\text{min})} = 0. \tag{17.1}$$

In a completely analogous manner we can save the classical law of conservation of mass by ascribing a proper mass $\Delta \mathscr{E}/c^2$ to the classical energy liberated and writing (16.19) in the form

$$\Delta M_{\text{cl}} = \Delta M^{(\text{min})} + \Delta \mathscr{E}/c^2 = 0, \tag{17.2}$$

so that we can talk of the conservation of the sum of proper masses plus the proper mass of the energy. Here, again, mass is converted into mass of a different type, but not into energy.

Thus, the assertion that mass is converted into energy is devoid of any physical meaning. The use of this assertion to obtain philosophical conclusions is also without any foundations.

Even the interpretation of deductions from (16.19) as the nonconservation of classical energy in transformations of matter cannot serve as the justification for the assertion that momentum is destructible, inasmuch as classical energy is only a part of the total energy which is conserved.

§ 18. THE NONADDITIVITY OF PROPER MASSES

In contrast to energy and momentum, proper mass is not an additive quantity, i.e., the proper mass of a system (M) does not equal the sum of the proper masses (M_k) of the particles forming it. This can be seen directly from the general expression for the proper mass of a system of particles which, according to (16.4), (16.5), is given by

$$M = \frac{1}{c} \sqrt{\left(\sum_k e_k / c\right)^2 - \left(\sum_k \mathbf{p}_k\right)^2}. \tag{18.1}$$

The proper masses of the individual particles are given by*

$$M_k = \frac{1}{c} \sqrt{(e_k / c)^2 - (\mathbf{p}_k)^2}. \tag{18.2}$$

---

*When a field is present, its mass and momentum should be included in the sums appearing on the right-hand side of (18.1).

Noting that, according to (14.13), the momenta can be expressed as

$$\mathbf{p} = \frac{e}{c}\boldsymbol{\beta}, \quad \text{where} \quad \boldsymbol{\beta} = \frac{\mathbf{u}}{c}, \tag{18.3}$$

we will write expression (18.1) as

$$M = \frac{1}{c}\sqrt{\left(\sum_k \frac{e_k}{c}\right)^2 - \left(\sum_k \frac{e_k}{c}\boldsymbol{\beta}_k\right)^2}. \tag{18.4}$$

It can be seen from this expression that $M = \sum_k M_k$ only if the velocities of all particles are equal, i.e.,

$$\boldsymbol{\beta}_k = \boldsymbol{\beta}. \tag{18.5}$$

If the particle velocities are different and all $e_k > 0$,* then $M > \sum_k M_k$, since, in the center-of-mass system, expression (18.4) has the form

$$M = \frac{1}{c^2}\sum_k e'_k = \sum_k \frac{M_k}{\sqrt{1-\beta_k'^2}} > \sum_k M_k, \tag{18.6}$$

where $\beta'_k = u'_k/c$, u'_k being the particle velocities in the center-of-mass system.

There is special interest in the case where the system consists of particles moving with the velocity of light. For such particles $\beta_k^2 = 1$ and, according to (18.2) and (18.3), their proper masses are equal to zero. However, formula (18.4) shows that the total proper mass of such a system consisting of particles with zero proper mass can be equal to zero only when condition (18.5) is satisfied, i.e., only when all the particles are moving in the same direction. When the directions of the particle velocities do not coincide, we have

$$\left(\sum_k \frac{e_k}{c}\boldsymbol{\beta}_k\right)^2 < \left(\sum_k \beta \frac{e_k}{c}\right)^2 = \left(\sum_k \frac{e_k}{c}\right)^2, \tag{18.7}$$

since $\beta_k^2 = \beta^2 = 1$. According to (18.4), this leads to M > 0.

---

*The question of the existence of particles for which $e_k < 0$ and $M_k < 0$ is discussed in detail in Chapter 7.

Thus, any spatially restricted light packet consisting of photons moving with the velocity of light, but with directions distributed within a finite solid angle will have a nonzero proper mass.

Hence, although we can say about individual photons that they have a zero proper mass, we cannot say the same about light in general. Any real light beam has a nonzero proper mass. Only an infinite-plane light wave, i.e., a beam of strictly collinear photons, has a total proper mass equal to zero. But this case of a light beam is almost never realized in practice, because any real light beam is spatially restricted, i.e., it is not an infinite-plane wave.

V.

ARE VELOCITIES HIGHER THAN THE VELOCITY OF LIGHT POSSIBLE?

§ 19. VELOCITIES OCCURRING IN PHYSICAL PROCESSES

In physics we deal with velocities less than or equal to the velocity of light, as well as with velocities exceeding the velocity of light. Macroscopic bodies and all known elementary particles with real* positive proper masses in all frames of reference move with velocities less than the velocity of light. Photons with proper mass zero move with the velocity of light. Physical processes propagating with a velocity above that of light are known to exist. For example, the phase velocity of de Broglie waves or the phase velocity of electromagnetic waves in rarefied plasma exceed the velocity of light. Thus, physical processes can be characterized by velocities that are less than c, as well as velocities that exceed c.

At the end of Section 13 we showed that a transformation from one inertial frame of reference to another cannot make sublight velocities into hyperlight velocities and, conversely, a hyperlight velocity cannot be transformed into a velocity less than that of light. On the other hand, processes propagating at the velocity of light have the same velocity c in all inertial frames of reference. In the derivation of this assertion, it was of course assumed that all inertial frames of reference have sublight velocities, i.e., $v < c$. This restriction is associated with the structure of the Lorentz transformations themselves, as they lose all of their meaning when $v > c$, because the primed coordinates become imaginary quantities, and this cannot be given any physical interpretation.

Thus, considering that physical space and time can only be represented by inertial frames of reference moving relatively to

---

*Note from translator: Real in the mathematical sense.

one another with velocities v < c, we come to the conclusion that there is a fundamental difference between processes propagating with sublight, light, and hyperlight velocities.

From a four-dimensional point of view, hyperlight velocities are described by a four-dimensional vector whose components are imaginary quantities, since, according to (13.2), we have for u > c,

$$(ds)^2 = (dx_0)^2(1 - u^2/c^2) < 0, \qquad (19.1)$$

i.e., ds is an imaginary quantity, so that the components U_k defined by (13.5) also can only be imaginary. Consequently, for the components of the hyperlight velocities we must replace (13.14) by

$$(U_0, \mathbf{U}) = \left(\frac{ic}{\sqrt{u^2/c^2 - 1}}, \quad \frac{i\mathbf{u}}{\sqrt{u^2/c^2 - 1}} \right), \qquad (19.2)$$

where u and $\sqrt{u^2/c^2 - 1}$ are real quantities. This expression shows also that even from a formal kinematic point of view hyperlight velocities are fundamentally different from sublight velocities and cannot be transformed into the latter by means of the Lorentz transformations.

As we have already noted above, the phase velocities of some types of waves may be higher than the velocity of light. As an example we will consider the waves of the spinor field ψ_α. In the case of a monochromatic plane wave*

$$\psi = \psi_0 e^{i[\omega t - \mathbf{kr}]} \qquad (19.3)$$

the phase velocity is

$$\mathbf{u} = \frac{\omega}{k} \mathbf{n}, \qquad (19.4)$$

where

$$\mathbf{n} = \mathbf{k}/k \qquad (19.5)$$

is the normal to the wave front or the direction of propagation of the wave. With the help of (19.4), the monochromatic wave ψ can

---

*The spinor index α is omitted in the following, and it is assumed that $\psi_1^* \psi_1 + \psi_2^* \psi_2 + \psi_3^* \psi_3 + \psi_0^* \psi_0 = \psi^* \psi$.

be written as

$$\psi = \psi_0 e^{i\omega(t-nr/u)}. \tag{19.6}$$

This expression shows that **u** determines the velocity with which a given surface of constant phase of a monochromatic plane wave propagates through space.

However, for any type of field, a monochromatic plane wave is an ideal particular solution of the field equation never realized in practice. In fact, we have to deal with wave packets restricted in space and time which can be represented as superpositions of plane waves with various amplitudes and wave vectors **k**, i.e., as

$$\psi(\mathbf{r}, t) = \int_{-\infty}^{+\infty} g(\mathbf{k}) e^{i[\omega(\mathbf{k})t - \mathbf{k}\mathbf{r}]} d\mathbf{k}, \tag{19.7}$$

where $d\mathbf{k} = dk_x dk_y dk_z$, while the function $\omega(\mathbf{k})$ governs the dispersion law for the waves under consideration.

Such a spatially restricted wave packet moves through space as a complete entity with a velocity that can be defined as the velocity of the "centroid" of the packet. By the coordinates of the "centroid" of the packet, we understand the components of the vector

$$\mathbf{R} = \frac{\int_{-\infty}^{+\infty} \mathbf{r} |\psi|^2 d\mathbf{r}}{\int_{-\infty}^{+\infty} |\psi|^2 d\mathbf{r}}. \tag{19.8}$$

The denominator of this expression in view of (19.7) can be written as

$$\iiint_{-\infty}^{+\infty} g(\mathbf{k}) g^*(\mathbf{k}') e^{i[\omega(\mathbf{k}) - \omega(\mathbf{k}')]t - i(\mathbf{k}-\mathbf{k}')\mathbf{r}} d\mathbf{k} \, d\mathbf{k}' \, d\mathbf{r} =$$

$$= (2\pi)^3 \int_{-\infty}^{+\infty} g(\mathbf{k}) g^*(\mathbf{k}) \, d\mathbf{k}, \tag{19.9}$$

since

$$\int_{-\infty}^{+\infty} e^{-i(\mathbf{k}-\mathbf{k}')\mathbf{r}} d\mathbf{r} = (2\pi)^3 \delta(\mathbf{k} - \mathbf{k}'), \tag{19.10}$$

where

$$\delta (k - k') = \delta (k_x - k'_x) \delta (k_y - k'_y) \delta (k_z - k'_z).$$

The numerator of (19.8) is equal to

$$\int\!\!\!\int\!\!\!\int_{-\infty}^{+\infty} g(k) g^*(k') r \, e^{i[\omega(k) - \omega(k')]t - i(k-k')r} \, dk \, dk' \, dr =$$

$$= \int\!\!\!\int\!\!\!\int_{-\infty}^{+\infty} g(k) g^*(k') e^{i[\omega(k) - \omega(k')]t} \, i\nabla_k e^{-i(k-k')r} \, dk \, dk' \, dr, \qquad (19.11)$$

where the operator ∇_k is a vector with the components

$$\nabla_k = \left(\frac{\partial}{\partial k_x}, \frac{\partial}{\partial k_y}, \frac{\partial}{\partial k_z} \right). \qquad (19.12)$$

If $g(k)$ is a function that decreases sufficiently rapidly as $k \to \infty$, then, integrating by parts and using (19.10), we obtain the following expression for the numerator of (19.8):

$$-(2\pi)^3 \int_{-\infty}^{+\infty} [g^*(k) i\nabla_k g(k) - g^*(k) g(k) t\nabla_k \omega(k)] \, dk. \qquad (19.13)$$

The denominator (19.9) is independent of time as is also the first term of the numerator (19.13). Consequently, the velocity of motion of the "centroid" is given by

$$\frac{dR}{dt} = V = \frac{\int_{-\infty}^{+\infty} \nabla_k \omega(k) |g(k)|^2 \, dk}{\int_{-\infty}^{+\infty} |g(k)|^2 \, dk}. \qquad (19.14)$$

The vector

$$\nabla_k \omega(k) = \left(\frac{\partial \omega}{\partial k_x}, \frac{\partial \omega}{\partial k_y}, \frac{\partial \omega}{\partial k_z} \right) = v \qquad (19.15)$$

is called the group velocity of the wave. Consequently, the velocity of the "centroid" of the packet can be considered, according to (19.14), as the average group velocity of the packet.

If the packet is quasi-monochromatic, i.e., $|g(k)|^2$ approaches a delta function of the type $\delta(k - k_0)$, then the average group velocity practically coincides with the true group velocity at $k = k_0$.

Consequently, the "centroid" of a quasi-monochromatic wave group moves with a group velocity $\nabla_k \omega(\mathbf{k}_0)$.

It is not difficult to see that the wave packet (19.7) not only moves as a whole, but also changes in width.

It is reasonable to take the measure of width to be the quantity $\Delta(\mathbf{r})$, defined in an analogous manner to the mean square deviation, by the expression

$$[\Delta(\mathbf{r})]^2 = \frac{\int\limits_{-\infty}^{+\infty} (\mathbf{r} - \mathbf{R})^2 |\psi|^2 \, d\mathbf{r}}{\int\limits_{-\infty}^{+\infty} |\psi|^2 \, d\mathbf{r}}, \qquad (19.16)$$

or

$$\Delta(\mathbf{r}) = \sqrt{\frac{\int\limits_{-\infty}^{+\infty} r^2 |\psi|^2 \, d\mathbf{r}}{\int\limits_{-\infty}^{+\infty} |\psi|^2 \, d\mathbf{r}} - R^2}. \qquad (19.17)$$

The numerator of the first term of (19.17) can be written with the help of (19.7) as

$$\int\limits_{-\infty}^{+\infty} r^2 |\psi|^2 \, d\mathbf{r} = \iiint\limits_{-\infty}^{+\infty} g(\mathbf{k}) \, g^*(\mathbf{k'}) \, r^2 e^{i[\omega(\mathbf{k}) - \omega(\mathbf{k'})]t - i(\mathbf{k} - \mathbf{k'})\mathbf{r}} \, d\mathbf{k} \, d\mathbf{k'} \, d\mathbf{r} =$$

$$= -\iiint\limits_{-\infty}^{+\infty} g(\mathbf{k}) \, g^*(\mathbf{k'}) \, e^{i[\omega(\mathbf{k}) - \omega(\mathbf{k'})]t} \, \nabla_k^2 e^{-i(\mathbf{k} - \mathbf{k'})\mathbf{r}} \, d\mathbf{k} \, d\mathbf{k'} \, d\mathbf{r}.$$

Integrating this expression by parts twice and using (19.10), we obtain

$$\int\limits_{-\infty}^{+\infty} r^2 |\psi|^2 \, d\mathbf{r} = -(2\pi)^3 \int\limits_{-\infty}^{+\infty} g^*(\mathbf{k}) \, \{\nabla_k^2 g(\mathbf{k'}) -$$

$$- it \, [2\nabla_k \omega(\mathbf{k}) \, \nabla_k g(\mathbf{k}) - g(\mathbf{k}) \nabla_k^2 \omega(\mathbf{k})] -$$

$$- t^2 g(\mathbf{k}) \, [\nabla_k \omega(\mathbf{k})]^2 \} \, d\mathbf{k}. \qquad (19.18)$$

Substituting (19.18) and (19.19) into (19.17) and, in addition, taking into account that according to (19.14)

$$\mathbf{R} = \mathbf{V}t + \mathbf{R}_0, \qquad (19.19)$$

we obtain

$$\Delta(\mathbf{r}) = A \sqrt{(t-t_0)^2 + D}, \qquad (19.20)$$

where

$$A^2 = \frac{\int\limits_{-\infty}^{+\infty} [\nabla_k \omega(\mathbf{k})]^2 \, |g(k)|^2 \, dk - \left[\int\limits_{-\infty}^{+\infty} \nabla_k \omega(\mathbf{k}) \, |g(k)|^2 \, dk\right]^2}{\int\limits_{-\infty}^{+\infty} |g(\mathbf{k})|^2 \, dk}. \qquad (19.21)$$

The constants t_0 and D can also be expressed in terms of V, R_0, and the integrals appearing in (19.18).

Thus, the rate of broadening of the wave packet tends to A as $t \to \infty$. This rate tends to zero for quasi-monochromatic packets as can be seen from (19.21).

We have considered here the case of a complex spinor field ψ. However, it can be shown that analogous results are also obtained for other types of fields and, hence, any quasi-monochromatic wave packet moves as a whole with the group velocity.*

Consequently, the motion of a wave packet is physically characterized by the group and not the phase velocity. It is obvious that the energy of the field also moves with this velocity as it is concentrated inside the packet, so-to-say at the "centroid."

It is found that for all known wave processes propagating with hyperlight phase velocities, the group velocity is less than the velocity of light. Thus, for example, in the case of de Broglie waves we have

$$\omega = c \sqrt{k^2 + c^2 M^2 / \hbar^2}, \qquad (19.22)$$

from which, according to (19.4) and (19.15), the product of the phase and group velocities is

$$uv = c^2. \qquad (19.23)$$

It follows from this that $v < c$ when $u > c$.

This confirms the widely held view that actual physical velocities with which energy is transmitted from place to place are

___

* This question has been examined by R. Serebryanyi [6].

sublight velocities or, at any rate, equal to the velocity of light. Only phase velocities are higher than the velocity of light, i.e., the velocities involved in the propagation of wave phases which are not associated with any mass or energy transfer.

On the other hand, as we will see below, this conclusion can be critically reviewed on the basis of a more detailed physical analysis.

§ 20. THE VELOCITY-OF-LIGHT LIMIT AS A CONSEQUENCE OF THE PRINCIPLE OF CAUSALITY

A general theorem is usually proved in the theory of relativity that no signal can propagate with a velocity higher than that of light. A "signal" is implicitly understood as a macroscopic interaction which can give rise to macroscopically observable events. It is also assumed that the signal carries the interaction from one point to another, i.e., it is emitted as desired at a point 1 and is absorbed at point 2 producing some macroscopic process.

Thus, the signal is represented as a macroscopic amount of energy which is carried by some physical agent from one point to another. The transmitting agent can be a particle with finite proper mass, an individual photon, or a wave group. The wave phase obviously cannot be a signal, since a plane wave must be distorted in some way in order for it to be able to carry a given discrete interaction. On the other hand, any distortion of a plane wave produces a wave group which will propagate at the group and not the phase velocity.

Let us prove the theorem mentioned above that signals with hyperlight velocities are impossible.

Let us consider two events: the emission of a signal at the point x_1 at time t_1 and the absorption of the signal at the point x_2 at a subsequent time t_2. The velocity of the signal is obviously given by

$$u = (x_2 - x_1) / (t_2 - t_1). \qquad (20.1)$$

It is assumed here that

$$t_2 > t_1, \qquad (20.2)$$

because the absorption of a signal is by definition a later event than its emission.

Let us now consider the same two events from the point of view of another inertial frame of reference Σ' which moves relatively to the first with a velocity v. According to the Lorentz transformations [see formula (11.6)], we have

$$t_2' - t_1' = \frac{(t_2 - t_1) - \frac{v}{c^2}(x_2 - x_1)}{\sqrt{1 - v^2/c^2}}, \qquad (20.3)$$

from which according to definition (20.1) we have

$$t_2' - t_1' = (t_2 - t_1)\frac{1 - vu/c^2}{\sqrt{1 - v^2/c^2}}. \qquad (20.4)$$

If u < c, then in view of the fact that we always have v < c, we find that $vu/c^2 < 1$, and the coefficient of $(t_2 - t_1)$ is a positive quantity. Consequently, $(t_2' - t_1') > 0$ if $(t_2 - t_1) > 0$, i.e.,

$$t_2' > t_1' \quad \text{when} \quad u < c. \qquad (20.5)$$

Thus, the temporal sequence of events associated with a sublight signal cannot be changed by means of a transition from one frame of reference to another.

The situation is different in the case of hyperlight signals. When u > c, then, according to (20.4), it is possible to select a frame of reference Σ' moving with a velocity v < c in which the coefficient of $(t_2 - t_1)$ is negative. Indeed, with u > c, i.e., c/u < 1, there exist such v < c for which we have

$$c/u < v/c < 1, \qquad (20.6)$$

which yields $vu/c^2 > 1$.

Thus, if the frame of reference has a velocity defined by inequality (20.6), then the coefficient of $(t_2 - t_1)$ in (20.4) becomes negative and, consequently, with $(t_2 - t_1) > 0$ we obtain $(t_2' - t_1') < 0$, i.e., the temporal sequence of events no longer holds, because with $t_1 < t_2$ we find that

$$t_2' < t_1'. \qquad (20.7)$$

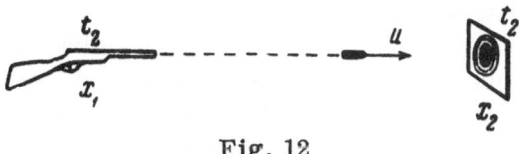

Fig. 12

Consequently, the admission of hyperlight signals is equivalent to the admission of the possibility that the temporal sequence of signal emission and absorption can be changed by a suitable choice of a frame of reference. However, the admission of this possibility contradicts the principle of causality in its formulation adopted in physics, since by a suitable choice of a frame of reference we can make the cause (signal emission) follow the effect (signal absorption).

The possibility of the reversal of the temporal sequence of cause and effect appears to be particularly absurd in the following macroscopic example.

Let a rifle be the source of the signal, which is here a bullet, while the receiver of the signal is a target (see Fig. 12). If the velocity of the bullet is less than the velocity of light, then from the view of any frame of reference, the firing of the rifle, i.e., the cause, always precedes the arrival of the bullet at the target, i.e., the effect. If the velocity of the bullet exceeds that of light, then we can always find a frame of reference in which the hitting of the target is an earlier event than the firing of the rifle.* The absurdity of this, i.e., the impossibility of such a reverse process ever being realized, is put forward as an obvious argument against the possibility of a breakdown of the principle of causality and, hence, as an argument that hyperlight signals are physically impossible.

---

*More accurately, this reversal of the sequence of events in time must be represented as a time reversal of the whole process, i.e., at first the thermal energy of the target becomes concentrated in such a manner that the bullet is ejected by the target and flies into the barrel of the rifle. At the same time, all of the combustion products of the propellant gather together in the one place and transform into the cartridge charge. The rifle becomes loaded and the process stops at this stage.

On the basis of this theorem, it is assumed in the theory of relativity that it is impossible for us to have bodies moving with a velocity higher than the velocity of light, and no processes can transmit energy or propagate with a velocity higher than that of light. It is also used to prove the impossibility of the existence of absolutely rigid bodies, i.e., elastic bodies in which elastic waves can propagate with an infinitely high velocity. In this context, these conclusions should not be considered as a consequence solely of the interconnection of space and time represented by the Lorentz group, but as a consequence of it and the Principle of Causality used in relativity theory. Moreover, the assertion that the theory of relativity, so to speak by itself, without recourse to other laws of nature, prohibits hyperlight signals and energy-transfer processes propagating with velocities higher than the velocity of light is incorrect.

Thus, we have been forced to make use of an additional assumption, called the "Principle of Causality," in order to make a purely physical deduction about the impossibility of the existence of hyperlight signals and other perturbations capable of inducing physical interactions. However, physical inferences can be arrived at only on the basis of physical laws and postulates. Thus, the Principle of Causality must be considered as a physical assertion or a physical law.

From the point of view of a general philosophical system, the Principle of Causality is associated with the notion that some events cause other events. This causal relationship is distinguished from a general relationship in that a direction is implied in the action, namely, the direction from cause to effect. The conditional relationship is taken to be a universal one, i.e., it is assumed that there are no phenomena in existence which are not conditional upon other phenomena. In this section, however, by the Principle of Causality we have meant a more restricted assertion concerning two events occurring at two spatially separated points x_1 and x_2 at times t_1 and t_2. These events we have considered to be causally related, the event at the point 1 being taken as the cause and the event at point 2 as the effect. Moreover, the Principle of Causality was interpreted as an assertion that the cause must necessarily precede the event, and that this temporal sequence of events is an absolute one, i.e., it cannot be affected by any choice of a frame of reference or the observer's point of view.

In other words, the Principle of Causality was taken to mean the following statement: Of two causally related events taking place at two spatially separated points, one is the cause, the other the effect, the cause always occurring before the effect, and this sequence cannot be changed by any choice of a frame of reference.

The essence of this assertion is that the temporal sequence of causally related events is absolute and cannot be disrupted or, speaking in everyday language, time flows in the one direction from the past into the future. Consequently, the "Principle of Causality" and the notion of the directivity of physical process in time are basically the same assumption [7, 8].*

§ 21. THE PRINCIPLE OF CAUSALITY AND
THE SECOND LAW OF THERMODYNAMICS

With the universal recognition of statistical physics as the basis of thermodynamics and, in general, the physics of macroscopic processes, it was established as early as the beginning of the present century that the directivity of physical processes is exclusively governed by statistical laws, a consequence of which is the second law of thermodynamics.

In view of the absolute reversibility in time of microscopic laws of motion, the direction in which time flows cannot be distinguished microscopically. All microscopic processes take place in a completely symmetrical fashion irrespective of whether they are considered in the positive (from past to future) or the negative (from future to past) direction in time. Any directivity in microscopic processes is due to the special initial conditions imposed by the macroscopic circumstances of an experiment. In other words, any irreversibility or directivity in time is a consequence of macroscopic irreversibility within the surrounding macrouniverse. Macroscopic irreversibility, in turn, is a consequence of the second law of thermodynamics which is purely statistical in nature.

Since the "Principle of Causality" in its narrow physical formulation is an expression of the directivity of processes in

---

*This point of view has also been investigated by Costa de Beauregard [19].

time, while the latter follows from the second law of thermodynamics, the "Principle of Causality" can be considered as a consequence or a particular expression of the second law of thermodynamics.

The alternative point of view, the rejection of the interrelation between the "Principle of Causality" and the second law would
imply an attempt to introduce a special physical law which postulates for all physical processes a definite and unvarying directivity in time. In the case of the microuniverse, this would mean
that an asymmetry in time is ascribed to the laws of micromotion.
This, however, has not yet been successfully introduced by anyone.
Thus, the intuitive wish to consider the "Principle of Causality"
as an absolute (i.e., inviolate everywhere) law of nature cannot be
justified on the basis of known laws of nature.

It is even difficult to conceive what would be the significance
of the discovery of a true irreversibility in the laws of micromotion. For any law described by differential equations, irreversibility in time would mean the appearance of terms which are non-
symmetric relative to a change in the sign of the time t. The latter is inevitably associated with a monotonic increase with time
of some physical quantity, i.e., with some type of nonconservation
law. On the other hand, everything in the microuniverse is built
on the conservation of momentum and the conservation of matter
with a great deal more certainty than in the macrouniverse.

Thus, there is every reason for believing that the "Principle
of Causality," i.e., the assertion of the invariance of the temporal
sequence of causally related, spatially separated events is merely
a rule of macroscopic origin which distinguishes a definite direction of flow of time. This rule is a simple consequence of the law
of entropy increase.

However, from this point of view, it is possible for the
"Principle of Causality" to be violated in the same way as violations of the second law of thermodynamics are possible in fluctuation processes. Consequently, the prohibitions imposed in the
theory of relativity by the Principle of Causality cannot be absolute ones and only involve processes of a macroscopic character.
Thus, it is essential to reexamine the conclusions that are made
on the basis of the "Principle of Causality" in the theory of relativity, namely, the conclusions concerning the impossibility of the

existence of signals or disturbances propagating with a velocity higher than the velocity of light.

So that we do not introduce any *a priori* elements arising from the intuitive concepts concerning the directivity of time into our discussion, we will proceed directly from the second law of thermodynamics and we will examine the prohibitions which it imposes on the transfer processes involving physical interactions.

§ 22. A SIGNAL AS A PHYSICAL PROCESS TRANSFERRING NEGENTROPY

From the point of view of information theory, the reception of a signal means the reception of information. Consequently, the sending of a signal from point 1 to point 2 means the transmission of information from the first point to the second.

It is known that information is proportional to negentropy, i.e., the amount by which the nonequilibrium entropy is in excess of the equilibrium entropy, so that by a signal we must understand a purely localized perturbation which, moving from the transmitter to the receiver, carries negentropy with it. The localized perturbation can transfer energy* and in this way change the equilibrium entropy of the receiver, without being a signal, if it does not transfer negentropy which is a measure of the degree of departure from equilibrium of the system.

The latter can be understood from a purely thermodynamic point of view without recourse to cybernetic arguments. Indeed, for the absorbed signal to be detected, it must initiate a macroscopic action in the receiver system, i.e., it must produce in it a spontaneous macroscopic process. The latter is only possible if negentropy is communicated to the system so that the system can be transferred from a metastable state into a stable state across the entropy minimum separating these states. Let ΔS_1 be the increase in the entropy of the radiator and ΔS_2 be the increase in the entropy of the signal receiver. In view of the second law of thermodynamics, it is obvious that

$$\Delta S_1 + \Delta S_2 \geqslant 0. \tag{22.1}$$

---

*By energy we mean its relativistic expression (14.13).

The process carrying an energy ΔE can be a signal only if

$$\Delta S_2 < \Delta S_2^0 \quad \text{or} \quad \Delta (S_2 - S_2^0) < 0, \tag{22.2}$$

where ΔS^0 is the equilibrium increase in the entropy of the system when its energy increases by an amount ΔE. It is clear that $\Delta S_1^0 = -\Delta S_2^0$, so that, in view of (22.1), we have

$$\Delta S_1 > \Delta S_1^0 \quad \text{or} \quad \Delta (S_1 - S_1^0) > 0, \tag{22.3}$$

i.e., when a signal is emitted, the negentropy of the transmitter $S_1^0 - S_1$ can only decrease.

Conditions (22.2) and (22.3), obviously, cannot be violated in a transition to another frame of reference if the signal is propagated with a velocity less than the velocity of light. However, in the case of hyperlight signals, it is possible to choose a frame of reference [see (20.6)] in which the absorption of a signal by the receiver will be an earlier event than its emission by the transmitter. In other words, the receiver has become the "transmitter" and the transmitter, the "receiver." However, in this case the emission of the signal will be accompanied by an increase in the negentropy of the "transmitter," and its absorption will involve a decrease in the negentropy of the "receiver," i.e., the "signal" will carry negentropy in the opposite direction.

According to what has been said above, such a process contradicts conditions (22.1)-(22.3). This process can lead to the violation of the second law of thermodynamics, because spontaneous emission with an increase in the negentropy of the transmitter can be used to construct a perpetuum mobile of the second kind. Hence, we see that hyperlight signals are indeed forbidden by the second law of thermodynamics. However, this prohibition is not an absolute one since fluctuations can violate the second law. Consequently, localized perturbations carrying entropy at hyperlight velocities are admitted as are any fluctuation processes.

This means that there can exist interactions propagating with hyperlight velocities, but they cannot be "signals" transferring negentropy. Such hyperlight processes cannot be excited by us at will at one point and absorbed at another distant point, because the systematic reproduction of such processes would mean the systematic (and not fluctuational) violation of the second law of thermodynamics.

The second law of thermodynamics forbids only macroscopic processes propagating with hyperlight velocities which can be used to transmit interactions capable of producing macroscopic irreversible processes. Among such processes, for example, is the process of the propagation of an elastic perturbation in a rigid body or, in other words, the process of sound propagation. Consequently, the velocity of sound in a rigid body cannot exceed the velocity of light. This means that there are no absolutely rigid macroscopic bodies, because the velocity of sound in an absolutely rigid body must be infinitely high, since an absolutely rigid body, by definition, cannot be deformed and must move only as a whole. With the help of an absolutely rigid body we could transmit macroscopic interactions instantaneously from one point to another point spatially distant from the first. This is, however, forbidden by the second law of thermodynamics and is therefore not realizable in the macrouniverse, i.e., absolutely rigid bodies cannot exist.

§ 23. ARE HYPERLIGHT INTERACTIONS POSSIBLE INSIDE ELEMENTARY PARTICLES?

The conclusion reached in the preceding section that absolutely rigid bodies do not exist can be extended without exception to all physical objects, in particular to elementary particles. It is asserted that elementary particles cannot have any extension, because elementarity denotes indivisibility, while only an object similar to an absolutely rigid body, in which perturbations are transmitted with hyperlight velocities, can be indivisible. Consequently, as stated by some authors, elementary particles must be mass points, since in the converse case it would be possible for processes propagating with hyperlight velocities to take place inside elementary particles, which is allegedly forbidden by the theorem proved above.

It is not difficult to see that this deduction is only made on the basis of a hypothesis that perturbations propagating inside elementary particles are in the nature of macroscopic signals, i.e., the processes occurring within elementary particles are subject to the "Principle of Causality" in its restricted formulation given at the end of Section 20, or that the second law of thermodynamics is of an absolute character inside elementary particles.

A more reasonable hypothesis, however, is the admission inside elementary particles of a very close connection between spatially separated events. This connection can be so strong that spatially separated events can only be considered to be causally interrelated and not separable into causes that occur at one point and effects that occur at another point. In this case a change in the temporal sequence of events does not lead to the violation of the causal interrelationship and the Principle of Causality in its wider sense. What is violated is merely a rule called the "Principle of Causality" which is a consequence of the law of increase of entropy. However, there are no grounds for us to require that the law of increase of entropy should hold for all processes taking place inside elementary particles. On the contrary, in view of the strictly macroscopic nature of the second law of thermodynamics, there is every reason to believe that the processes inside elementary particles occur predominantly as fluctuation processes, i.e., they take place with the violation of the second law of thermodynamics.

Thus, a broader approach to the content of the theory of relativity and the Principle of Causality removes the necessity for elementary particles to be point masses. With this approach, elementary particles can have internal structure, physical processes being propagated inside the particles with hyperlight velocities.

The hypothesis of the violation of the Principle of Causality at distances less than 10^{-13} cm has been proposed in the past (Blokhintsev, Heisenberg, and others). However, a serious objection has been put forward, namely, that the violation of the Principle of Causality at distances of the order of 10^{-13} cm by a suitable choice of a reference system can be transformed into the violation of the Principle of Causality over an arbitrarily large distance, i.e., the admission of the violation of the Principle of Causality in the microuniverse also leads to its violation in the macrouniverse.

Let us examine this proof more closely. Let two interrelated events x_1, t_1 and x_2, t_2 be separated by a space-like interval, i.e.,

$$s^2{}_{12} = c^2 (t_2 - t_1)^2 - (x_2 - x_1)^2 < 0 \qquad (23.1)$$

or, what is the same,

$$u = \frac{(x_2 - x_1)}{(t_2 - t_1)} > c. \tag{23.2}$$

In another frame of reference according to the Lorentz transformations we have

$$x'_2 - x'_1 = \frac{(x_2 - x_1) - v(t_2 - t_1)}{\sqrt{1 - v^2/c^2}} = (x_2 - x_1)\frac{1 - v/u}{\sqrt{1 - v^2/c^2}}. \tag{23.3}$$

According to this formula, with $u \gg c$, we obviously have

$$x'_2 - x'_1 = \frac{x_2 - x_1}{\sqrt{1 - v^2/c^2}}. \tag{23.4}$$

Consequently, if a process takes place between the points x_1 and x_2 violating the "Principle of Causality," then in system Σ' this process takes place between the points x'_1 and x'_2 where the distance $l' = x'_2 - x'_1$ can be made arbitrarily larger than $l = x_2 - x_1$ as $v \to c$.

Let us note, however, that the probability of violation of the Principle of Causality decreases by the same factor as the distance between the events is increased. Indeed, if the "causeless" events occur in system Σ with an average frequency ν, then in the system Σ', in view of the retardation of clocks, the same events will occur with a frequency

$$\nu' = \nu \sqrt{1 - v^2/c^2}. \tag{23.5}$$

Consequently, we find that

$$l'\nu' = l\nu, \tag{23.6}$$

i.e., if the distance between "causeless" events increases by a factor N, then the frequency of occurrence of these events decreases by the same factor. This means that the probability of encountering such events also decreases by the factor N.

Thus, although the violation of causality can be made into a macroscopic effect, the probability of such a violation becomes vanishingly small. This is in good agreement with the thermodynamic interpretation of the "Principle of Causality," since from the statistical point of view, arbitrarily large violations of the second law of thermodynamics are possible, but the probability of such fluctuations decreases as the scale of the violations increases.

Thus, with the thermodynamic interpretation of the "Principle of Causality," the transfer of energy from one point to another with a hyperlight velocity is not absolutely forbidden, but is admitted as a fluctuation process.

Particles moving with hyperlight velocities also become physically admissible entities, but they cannot be arbitrarily emitted and absorbed to excite irreversible macroscopic processes.

According to (14.3) and (14.7), for any particle we have

$$\mathbf{P} = \frac{E}{c^2}\, \mathbf{u}, \quad c^2 M^2 = E^2/c^2 - \mathbf{F}^2; \qquad (23.7)$$

therefore, for particles moving with superlight velocities $u > c$ we find that

$$M^2 < 0, \qquad (23.8)$$

i.e., their proper mass is an imaginary quantity.

Consequently, we have come to the conclusion that it is physically admissible for particles to exist with an imaginary proper mass and move with velocities higher than the velocity of light.

VI.
NEGATIVE AND IMAGINARY PROPER MASSES

§ 24. DEFINITION OF PROPER MASS

At the end of the preceding paragraph it was shown that the framework of the theory of relativity admits particles moving with velocities higher than the velocity of light. Such particles are only prohibited by the Principle of Causality. However, the prohibition is removed if the Principle of Causality is not considered as an absolute physical law, but as a consequence of the second law of thermodynamics [9, 10].

The theory of relativity also admits particles of imaginary mass. This was first demonstrated by Dirac [11] in connection with the quantum theory of the electron. Such particles are commonly taken to be physically inadmissible on the basis of additional arguments concerning the nonviolation of postulates of thermodynamics and the Principle of Causality, i.e., on the basis of the same considerations that are used to show that hyperlight particles of imaginary mass are forbidden.

However, it was shown in Chapter V that the additional arguments derived from the Principle of Causality or thermodynamics cannot be considered to lead to an absolute prohibition and, therefore, the hypothesis of the existence of particles with negative and imaginary masses can be considered to be sound.

In order to clarify the implications of this hypothesis, we will examine the consequences following from it. First of all, we will study the general properties of particles with negative and imaginary masses on the basis of relativistic dynamics.

In accordance with (14.17), proper mass M is defined as the invariant length of the four-dimensional momentum vector P_k, i.e., by the relation

$$c^2 M^2 = P_\alpha P_\alpha = \frac{E^2}{c^2} - \mathbf{P}^2. \qquad (24.1)$$

It should be noted that in the following, for brevity, we will frequently use the term "mass" for the term "proper mass."

For a system of particles, we obviously have

$$E = cP_0 = \sum_k e_k, \quad \mathbf{P} = \sum_k \mathbf{p}_k, \tag{24.2}$$

where e_k and $\mathbf{p}_k$ are the energy and momentum of an individual particle whose mass is defined by $c^2 m_k^2 = e_k^2 - \mathbf{p}_k^2$. In the usual case, when $M^2 > 0$, the sign of the mass M is the same as that of the energy E.

According to (24.1), the proper mass is a real quantity if the vector P_k is time-like (i.e., $M^2 > 0$) and an imaginary quantity if this vector is space-like (i.e., $M^2 < 0$). The case of zero proper mass (M = 0) can be considered as a special case of real proper mass.

Thus, if the components of the vector P_k are taken to be arbitrary real numbers, then formula (24.1) admits of three essentially different physical systems:

1. systems with positive proper mass, i.e., $M^2 \geq 0$, $E > 0$;

2. systems with negative proper mass, i.e., $M^2 \geq 0$, $E < 0$;

3. systems with an imaginary proper mass, i.e., $M^2 < 0$.

Consequently, the framework of the theory of relativity admits three types of essentially different systems of which only systems of the first kind are considered to be physically real. Systems of the second and third kinds are forbidden by the Principle of Causality or the propositions of thermodynamics. Systems of the second kind obviously include the Dirac antielectrons of negative mass, and systems of the third kind include the virtual particles of quantum theory of field. Both of these are considered to be without a physical existence. However, within the framework of relativistic kinematics and dynamics there are no grounds for excluding these particles. The Principle of Causality, on the other hand, is of a different macroscopic nature and does not follow from relativistic kinematics or dynamics.

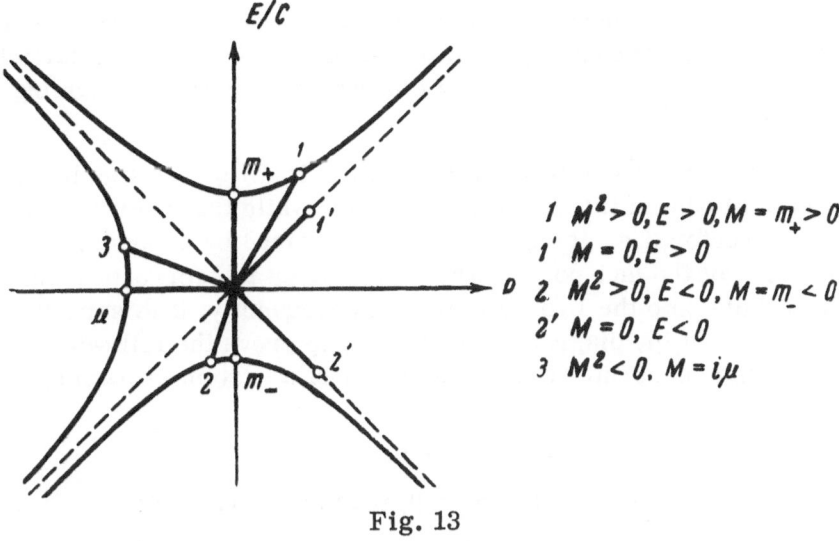

$$1 \quad M^2 > 0, E > 0, M = m_+ > 0$$
$$1' \quad M = 0, E > 0$$
$$2 \quad M^2 > 0, E < 0, M = m_- < 0$$
$$2' \quad M = 0, E < 0$$
$$3 \quad M^2 < 0, M = i\mu$$

Fig. 13

§ 25. THE GENERAL PROPERTIES OF PARTICLE SYSTEMS WITH POSITIVE, NEGATIVE, AND IMAGINARY MASSES

It is convenient to use the (P, E/c) diagram shown in Fig. 13 for the study of the general properties of all the three kinds of particle. The figure shows particles of all three types, including two subclasses of particles with zero proper masses and positive (1) and negative (2) energies. The absolute magnitudes of the masses m_+, m_-, and μ are found from the points of intersection of the E/c axis of the P hyperbolas with lines drawn through the points representing the corresponding particles.

In order to provide a more complete description of these systems, we will introduce the concept of the average velocity of a system

$$V = \frac{P}{E}c^2. \tag{25.1}$$

It is clear that $|V| \leq c$ for systems of the first and second kinds, and $|V| > c$ for particles of the third type. If the system consists of one particle, then the average velocity V coincides with the true velocity of the particle v.

Individual possible-particles can obviously be classified in the same manner as systems, but the class of a group of particles may not be the same as that of the individual particles forming this system.

In view of (24.2), the total vector $(\mathbf{P}, E/c)$ is a simple geometrical sum of the vectors $(\mathbf{p}_k, e_k/c)$, while the invariant length of the vector (i.e., the quantity M) is obtained by means of a "projection" of the end-point of the vector along the corresponding hyperbola onto the E/c or P axis. Consequently, it is sufficient to make use of the diagram (see Fig. 13) to prove the following theorems concerning the proper mass of a system of particles:

I. If all $m_k > 0$, then $M > 0$;

II. if all $m_k = 0$, but $e_k > 0$, then $M \geq 0$ $(E > 0)$;

III. if all $m_k < 0$, then $M < 0$;

IV. if all $m_k = 0$, but $e_k < 0$, then $M \leq 0$ $(E < 0)$;

V. if all $m_k^2 < 0$, then M remains arbitrary $(M^2 \geq 0, M^2 < 0)$.

In other words, a group of particles of the first kind can only form a system of the first kind, a group of particles of the second kind, a system of the second kind, and a group of particles of the third kind, a system of any kind.

In an analogous manner we can prove the following theorems for systems containing particles of different kinds:

VI. A group of particles containing more than one particle of the first kind and more than one particle of the second kind can form a system of any kind;

VII. a group consisting of one particle with positive masses m_+ and one particle with negative mass m_- can form a system with total mass $M > 0$ or $M^2 < 0$ (i.e., of the first or third kinds) if $m_+ + m_- > 0$; with total mass $M < 0$ or $M^2 < 0$ (i.e., of the second or third kinds) if $m_+ + m_- < 0$, and with total mass $M = 0$ or $M^2 < 0$ if $m_+ + m_- = 0$;

VIII. a group of any number of particles of the first and third kinds can only form a system of the first or third kind and, similarly, particles of the second and third kind can only form a system of the second or third kind.

The theorems given above illustrate the close connection existing between particles of positive, negative, and imaginary masses.

All known experimentally detected particles belong to the first class, which includes subclass 1' (photons), i.e., they have $M^2 \geq 0$ with $E > 0$. Systems of such particles also belong to this class according to I and II. However, it is sufficient to admit the existence of even two particles of the second type with negative proper masses for us to be forced to admit, in accordance with VI, the existence of all three classes of proper mass. Consequently, in introducing negative masses, we thereby also introduce imaginary proper masses.

Isolated particles with positive proper masses according to (25.1) can only have velocities less than the velocity of light, particles of zero mass, only velocities equal to the velocity of light, and particles of imaginary mass, only velocities higher than the velocity of light. However, how are we to interpret the deduction from (25.1) and VI that collections of particles of positive and negative mass can have a total imaginary proper mass and, consequently, an average velocity higher than the velocity of light? For each of the particles forming the system moves with a velocity less than that of light.

In order to clarify this situation, let us consider a system consisting of one particle with positive mass and one particle with negative mass, where $m_+ = |m_-| = m$. Let these particles move along the x axis with equal speeds, but in opposite directions. For such a system we obviously have $E = 0$, $P = 2mv$, and, consequently, $M^2 = -4m^2v^2 < 0$, $V = \pm\infty$. Let us assume that both particles are simultaneously absorbed by bodies A and B situated at the points $x = +a$ and $x = -a$. In view of the fact that the absorption of a particle of positive mass leads to an increase in the energy of body A, whereas the absorption of a particle of negative mass decreases the energy of body B, this process is equivalent to the emission of an amount of energy by B and the absorption of this energy by A. Since both these processes occur simultaneously, the whole process can be considered as the transfer of energy from body B to body A at an infinite velocity. By a suitable choice of a system of reference this velocity may be made finite, although in any system it will exceed the velocity of light in accordance with the theorem proved by us at the end of Section 13.

Thus, the average velocity V can be interpreted as the velocity with which energy, to be understood relativistically, is transferred from point to point. It is clear, however, that the process considered above cannot be used for signal transmission (i.e., information transmission) from point B to point A, because the events A and B cannot be in a cause—effect relation. Consequently, the signal cannot be transmitted with a velocity higher than that of light, although it is possible to have energy transfer at this velocity.*

§ 26. IS THERE A VIOLATION OF THE LAWS OF CONSERVATION OF ENERGY AND MOMENTUM?

In accordance with the Lorentz transformations and the formulas for the transformation of four-dimensional velocity (13.8) following from them, the components of four-dimensional momentum transform as

$$P'_1 = \frac{P_1 - \beta P_0}{\sqrt{1 - \beta^2}}, \quad P'_2 = P_2, \quad P'_3 = P_3, \quad P'_0 = \frac{P_0 - \beta P_1}{\sqrt{1 - \beta^2}}, \tag{26.1}$$

where $\beta = v/c$. For a particle moving along the x axis, we have

$$P' = P \frac{1 - v/u}{\sqrt{1 - v^2/c^2}}, \quad E' = E \frac{1 - vu/c^2}{\sqrt{1 - v^2/c^2}}, \tag{26.2}$$

where

$$u = c^2 \frac{P}{E} \tag{26.3}$$

is the velocity of the particle.

---

*We are giving no consideration here to the quantum properties of particles of negative mass. Attempts to take these into consideration have been made by Tanaka [22] and Feinberg [23].

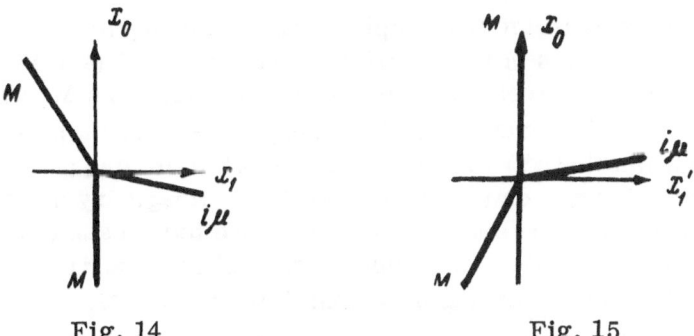

Fig. 14   Fig. 15

According to these formulas, in the case of particles of the first and second kinds energy does not change sign, while the momentum can change sign in a transformation to another frame of reference, since for such particles u < c. On the other hand, in the case of particles of the third type, u > c and, therefore, the momentum is of the same sign in all frames of reference, while the energy changes sign when the relative velocity v reaches the value

$$v = c^2/u. \qquad (26.4)$$

In this case, the three-dimensional velocity measured in the system Σ' also changes sign as can be seen from the velocity-transformation formulas (13.16), according to which

$$u' = \frac{u - v}{1 - vu/c^2}. \qquad (26.5)$$

Thus, in the case of hyperlight particles, the direction of the three-dimensional velocity and the direction of the three-dimensional momentum defined by the transformation formulas (26.2) do not coincide in all frames of reference. But this means that formulas (26.1), (26.2) lead to a paradox because the objective fact of the coincidence or noncoincidence of these directions depends on the choice of the frame of reference, i.e., on the method of representing space and time chosen by the physicist. The latter becomes even more paradoxical if we note that formulas (26.2) lead to the nonconservation of energy and momentum in processes of emission and absorption of hyperlight particles in some frames of reference.

Let us consider the simplest case of absorption of a particle of imaginary mass $i\mu$ by a particle of real mass M (see Fig. 14). It is known that a free particle of constant real mass M cannot absorb or emit another particle of real mass. Such a process, however, is allowed by the formulas of relativistic dynamics for the absorption and emission of particles of imaginary mass. Indeed, because of the conservation laws, the total mass of the system M is conserved. Consequently, for colliding particles of masses M_1 and M_2 turning into a particle of mass M, we have

$$c^2 M^2 = \left(\frac{E_1}{c} + \frac{E_2}{c}\right)^2 - (\mathbf{P}_1 + \mathbf{P}_2)^2 = c^2 M_1^2 + c^2 M_2^2 + 2\left(\frac{E_1}{c}\frac{E_2}{c} - \mathbf{P}_1\mathbf{P}_2\right). \quad (26.6)$$

If the particle M_1 was at rest before the collision, then $P_1 = 0$, $E_1 = M_1 c^2$, which according to (26.6) yields

$$E_2 = -\frac{c^2}{2M_1}[M_2^2 + M_1^2 - M^2]. \quad (26.7)$$

In the process under consideration, the first particle of real mass will not change its proper mass as the result of an absorption of the second particle of imaginary mass, i.e., $M_1 = M$, which yields

$$E_2 = -\frac{c^2}{2M}M_2^2. \quad (26.8)$$

On the other hand, we have for this process

$$E_1 + E_2 = E > E_1 = Mc^2, \quad (26.9)$$

where E is the energy of the first particle after it has absorbed the second. Consequently, we find that

$$E_2 > 0. \quad (26.10)$$

But the latter inequality can only be satisfied according to (26.8), when $M_2^2 < 0$, i.e., for a particle of imaginary mass. On the other hand, in the case of a particle of real mass, this inequality can only be satisfied when

$$M^2 > M_1^2 + M_2^2 > M_1^2, \quad (26.11)$$

i.e., for $M_1 \neq M$.

Let us consider the same process in another coordinate system in which the first particle (of real mass) is at rest after it interacts with the second particle (of imaginary mass) (see Fig. 15). In this frame of reference, the particle of imaginary mass $i\mu$ is now not absorbed by the particle of real mass, but emitted by it. Therefore, we must replace inequality (26.9) by

$$E_1 = Mc^0 + E_2' > Mc^2, \tag{26.12}$$

where E_2' is the energy of the emitted particle of imaginary mass. Consequently we have

$$E_2' > 0. \tag{26.13}$$

Hence, because of the law of conservation of energy the energy of the particle of imaginary mass must be positive in both frames of reference. However, in view of formulas (26.2) and (26.5), we have

$$E_2' = - E_2, \tag{26.14}$$

inasmuch as

$$u_2' = - u_2, \tag{26.15}$$

as is obvious from Figs. 14 and 15. But (26.14) clearly contradicts inequalities (26.10) and (26.13) obtained from the law of conservation of energy. Consequently, the transformations (26.2) applied to hyperlight particles lead to an expression which contradicts the law of conservation of energy when the three-dimensional velocity changes sign as the result of these transformations. It is not difficult to show that the law of conservation of momentum is also violated in this case. This can be seen from a comparison of Figs. 14 and 15. In Fig. 14 the momentum of the particle of imaginary mass is directed toward the left. Whereas in Fig. 15 the momentum of this particle points to the right. But in accordance with formulas (26.2), the momentum does not change sign and, consequently, these formulas in the case under consideration are not compatible with the law of conservation of momentum.

This contradiction between the formulas for the transformation of the components of four-dimensional momentum and the conservation laws can be removed if the right-hand side of formulas (26.2) are multiplied by minus one in those cases when

$$vu > c^2. \tag{26.16}$$

In other words, the formulas that are compatible with the laws of conservation of energy and momentum are not (26.2), but

$$P' = u' \frac{E'}{c^2}, \quad E' = E \frac{|1 - vu/c^2|}{\sqrt{1 - v^2/c^2}}, \quad u' = \frac{u - v}{1 - vu/c^2}. \quad (26.17)$$

These formulas clearly coincide with formulas (26.2) when $vu < c^2$. Consequently, formulas (26.17) are identical with (26.2) for any particles of real mass, because in this case $u < c$. Formulas (26.17) can also be written in a more general form analogously to formulas (26.1), namely,

$$P'_1 = \gamma \frac{P_1 - \beta P_0}{\sqrt{1 - \beta^2}}, \quad P'_2 = \gamma P_2, \quad P'_3 = \gamma P_3,$$

$$P'_0 = \gamma \frac{P_0 - \beta P_1}{\sqrt{1 - \beta^2}}, \quad \gamma = \frac{1 - \beta \frac{P_1}{P_0}}{\left| 1 - \beta \frac{P_1}{P_0} \right|}. \quad (26.18)$$

It is clear that for hyperlight particles, the four-dimensional momentum should be defined by formulas (26.17) and (26.18) because only in this case will the laws of conservation of energy and momentum remain inviolate in transformations to a new frame of reference satisfying condition (26.16).

In accordance with formulas (26.17) and (26.18), the energy of a particle of the third type does not change sign under all Lorentz transformations. Consequently, the particles of the third type can be subdivided into two subclasses, one containing particles with positive energy, the other particles with negative energy. The particles belonging to one subclass cannot be changed into particles of the other subclass by means of Lorentz transformations [24].

§ 27. PARTICLES OF NEGATIVE MASS
IN A GRAVITATIONAL FIELD

The motion of particles of negative mass is clearly governed by the Minkowski equations (14.1)-(14.4). Consequently, the acceleration of particles of negative mass is in a direction opposite to that of the impressed force F_k, since, in accordance with (14.1),

$$\frac{d^2 x_k}{d\tau^2} = \frac{dU_k}{d\tau} = \frac{F_k}{M}. \quad (27.1)$$

Thus, in an external electric field, identically charged particles
of positive and negative mass will be accelerated in opposite directions. However, in an external gravitational field, minus particles* must be accelerated in the same direction as the plus particles, since otherwise the minus particles would violate the principle of equivalence according to which all objects, without exception, acquire the same acceleration in the same gravitational field.
Consequently, on the basis of the principle of equivalence, we must
assume that in an external gravitational field minus particles of
mass ⌐m are subject to a force ⌐F, equal in absolute magnitude,
but opposite in sign to the force F acting on a plus particle of
mass m. In other words, we must assume that minus particles,
in the same way as plus particles, possess a "gravitational
charge"† given by one-and-the-same expression

$$ e = \sqrt{-\varkappa}\, M, \tag{27.2} $$

where $\varkappa$ is the gravitational constant.

Hence the usual large gravitating masses (the earth, sun,
stars, galaxies, etc.) attract minus particles as well as plus particles, imparting to them the same acceleration.

The situation is different if we consider the effects of large
negative masses on the behavior of plus and minus particles. They
will be repelled by these masses, inasmuch as the gravitational
field of negative masses is in the opposite direction to that of positive masses, since it can be considered to be produced by a "gravitational charge" of opposite sign.

Obviously, the gravitational interaction between minus particles forming a system will give rise to an effect different from
the effect of the gravitational interaction between particles of

---

* For brevity, in the following we will use the term plus particles
  for the usual particles of positive mass and the term minus particles for the particles of negative mass.
† In the nonrelativistic region, the gravitational field is governed
  by Newton's law of gravitation and can be considered, by analogy
  with the electric field governed by Coulomb's law, as being produced by a "gravitational charge" defined by Eq. (27.2).

positive mass. Indeed, gravitational forces draw plus particles together. But minus particles as the result of gravitational interactions will repel one another. Hence, condensations of matter should be formed in space filled with plus particles and matter will be distributed with a highly nonuniform density, as is the case in the real universe filled with plus particles. If space is filled with minus particles, the repelling forces will lead to a particle distribution of uniform density.

Assuming the existence of minus particles, we can consider a model of the universe with zero average proper mass. In the real universe, the average mass density due to plus particles is extremely low, being about 10^{-30} g/cm^3, i.e., approximately one proton per cubic meter. Moreover, intergalactic space is filled by a medium with a somewhat lower density, whereas higher densities, reaching 10^5 and even 10^9 g/cm^3 occur only inside stars, which occupy a negligible fraction of the total volume of cosmic space. If it is assumed that minus particles exist side by side with plus particles in the universe and that their mean concentration is approximately the same as that of plus particles, then, in view of the mutual repulsion of minus particles, we must postulate that the concentration of minus particles at any point in the universe is approximately constant. However, this concentration is very low, and this may be the explanation of the circumstance that we do not detect minus particles under terrestrial conditions, and do not detect their interactions in the surrounding medium consisting of plus particles. On the other hand, the presence of a background of minus particles could lead to appreciable effects on a galactic scale. Thus, for example, the minus particles are distributed with an approximately constant density ρ, and they must therefore completely cancel out the gravitational field of an individual galaxy of mass M at distances of the order of

$$R = (3M/4\pi\rho)^{1/3}, \tag{27.3}$$

since the absolute magnitude of the mass of minus particles contained in a sphere of radius R is equal to the mass of the galaxy situated inside this sphere.

We can make a more detailed estimate of the degree of gravitational screening if we assume that minus particles are in a state of equilibrium at a negative temperature $T < 0$ and are distributed in the field of a large plus mass according to Boltzmann's law

$$n^- = n_0^- e^{-m^-\varphi/kT}, \tag{27.4}$$

where n^- is the concentration of minus particles, n_0^- is the average concentration of minus particles in the metagalaxy, m^- the mass at minus particles, and φ the potential of the gravitational field defined, in the Newtonian approximation, by the equation

$$\nabla^2\varphi = 4\pi\varkappa\,(m^+ n^+ + m^- n^-). \tag{27.5}$$

For a universe with zero average mass, we obviously have

$$m^+ n_0^+ + m^- n_0^- = 0. \tag{27.6}$$

Noting further that the density due to free plus particles creating the gravitational field is

$$\rho = m^+\,(n^+ - n_0^+), \tag{27.7}$$

and assuming that

$$m^-\varphi/kT \ll 1, \tag{27.8}$$

we find for φ the following equation:

$$\nabla^2\varphi - \frac{1}{\Lambda^2}\,\varphi = 4\pi\varkappa\rho, \tag{27.9}$$

where

$$\Lambda = \sqrt{\frac{k\,|T|}{4\pi\varkappa\,|\,m^-\,|^2\,n_0^-}}\,. \tag{27.10}$$

Hence, instead of Poisson's equation, the Newtonian gravitational potential is described by Neumann's equation with a cosmological constant Λ depending on the temperature of minus particles and their mean concentration.

At distances exceeding Λ, the gravitational field of any body consisting of plus particles is completely screened. Individual galaxies with dimensions smaller than Λ will not be attracted to one another at distances exceeding Λ. On the other hand, if a galaxy has dimensions exceeding Λ, then its peripheral parts will not be retained by the gravitational attraction of the central region, so that ejection of matter from such a galaxy should occur.

Assuming that $n_0^- = 10^{-6}$ cm^{-3}, T = 3°K , we find for minus particles whose mass is equal to the nucleon mass that according to formula (27.10), $\Lambda = 1.3 \cdot 10^{22}$ cm.

It should be noted that a model of the universe with zero average mass density is of particular interest in relativistic cosmology.

§ 28. NEGATIVE MASSES AND THE PRINCIPLE OF CAUSALITY

Of all known objections to the existence of particles of negative mass, the most serious one is the objection based on the Principle of Causality or, equivalently, on the basis of the second law of thermodynamics (see [9, 10]). All other objections can either be reduced to it, or can be shown to be linked with it to some degree or other. Let us examine this principal objection.

Let us assume that particles of negative mass can be emitted or absorbed by systems of usual particles in the same way as, for example, photons or π^0 mesons. Let us suppose that a minus particle emitted by system A at time t_1 is absorbed by system B at time t_2, where $t_2 > t_1$. However, the emission of a minus particle means an increase in the energy and momentum of system A by amounts which are exactly the same as would be produced by the absorption of a plus particle of the same mass (in absolute magnitude) and velocity. Similarly, the absorption of a minus particle is equivalent to the emission of a plus particle by the system.

Thus, from the macroscopic point of view, the emission of a minus particle by system A and its absorption by system B is equivalent to the process of emission of a plus particle by system B and its absorption by system A, where the absorption in the latter process occurs at an earlier time than the emission, i.e., the Principle of Causality is violated.

From a thermodynamic point of view, this means that the process takes place with the violation of the second law of thermodynamics. Indeed, the emission of a plus particle is a process accompanied by a decrease in the negentropy of the emitter. Moreover, the particles emitted carry negentropy (i.e., positive in-

formation), while the negentropy of the absorbing system increases
at the moment of absorption of a plus particle (i.e., the system
goes into an excited state as a result of which irreversible processes can arise in it). On the other hand, the emission of a
minus particle is obviously a process accompanied by an increase
in the negentropy of the emitter, whereas their absorption leads
to a decrease in the negentropy of the absorber and minus particles
must therefore carry negative information (i.e., negative negentropy). But emission precedes absorption, i.e., the increase in
negentropy of the emitting body occurs spontaneously (the body appears to excite itself spontaneously), i.e., we have here a process
which violates the second law of thermodynamics. Thus, the
physically ill-defined assertion of the violation of the Principle of
Causality is replaced by the assertion that minus particles violate
a quantitatively well-defined physical law — the second law of
thermodynamics.

Usually, on the basis of arguments analogous to those given
above, it is categorically asserted that particles of negative mass
cannot exist.* This prohibition appears to be particularly "obvious" when the argument is based on the Principle of Causality and
not the second law of thermodynamics. If, however, it is recog-
nized that the Principle of Causality is merely an intuitively evi-
dent expression of the second law of thermodynamics, then the "ob-
viousness" of the prohibition of particles of negative mass loses
all physical justification.

Indeed, the second law of thermodynamics is a statistical
law reigning in the macroscopic world surrounding us, while de-
partures from it in individual fluctuations are allowed by statistical mechanics. Thus, events which violate the second law of
thermodynamics are admissible in individual fluctuations. Consequently, the second law of thermodynamics cannot forbid the existence of minus particles provided that for some reason they only
manifest themselves in processes of the fluctuation type.

Hence, a reference to the Principle of Causality cannot be
considered to be an argument in favor of an absolute prohibition

---

* The impossibility of the existence of particles of imaginary mass
 is proved in the same manner.

of minus particles. However, it is essential to establish whether
the minus particles violate the second law of thermodynamics,
not only in fluctuations, but also in the large and, more important,
if such violations do take place, can they be justified on physical
grounds?

§ 29. NEGATIVE MASSES AND THE SECOND LAW OF THERMODYNAMICS

In admitting the existence of particles of negative mass, we
assume that physical systems can possess arbitrarily large positive, as well as arbitrarily large negative, energies, there being
no lower limit to the negative energies. This property of systems
containing minus particles, however, contradicts one of the fundamental axioms of thermodynamics — the postulate that a state of
thermodynamic equilibrium exists. According to this postulate,
any thermodynamic system must have a state of thermodynamic
equilibrium, i.e., a state in which all macroscopic parameters are
constant and the system cannot spontaneously leave this state,
i.e., in the absence of any external interactions. However, not all
physical systems possess such a state of equilibrium. In thermodynamics and statistical physics we normally deal with systems
whose energy spectrum is bounded from below, i.e., there is a
minimum energy which is attained at zero temperature. Such systems do have a state of thermodynamic equilibrium. On the other
hand, in accordance with the laws of thermodynamics, systems
without a lower bound to their energies will undergo transitions to
lower and lower energy levels, i.e., they will be in a nonequilibrium state all the time. According to statistical physics, systems
with unbounded energy spectra, similar to systems containing
minus particles, are also without a state of thermodynamic equilibrium, inasmuch as they do not possess a convergent sum over
states,

$$Z = \sum_k e^{-E_k/\Theta}, \tag{29.1}$$

where $\Theta = kT$ is the modulus of the canonical distribution; T being
the absolute temperature; k, Boltzmann's constant; and E_k, the energy level of the system.

Thus, in admitting the existence of minus particles, we thereby concur with the possibility that the laws of thermodynamics can be violated not only in fluctuations, but also on a macroscopic scale. Of course, on the basis of the concept of the absolute inviolability of thermodynamic laws, we can consider the inconsistency between the existence of minus particles and the postulate of the existence of a thermodynamic equilibrium as an argument which proves that minus particles cannot exist. However, the atomistic viewpoint which underlies statistical physics is a more general approach, according to which the laws of thermodynamics are not absolute in character, but are merely consequences of the statistical theory of dynamic systems. According to this viewpoint, violations of the laws of thermodynamics are admissible not only in fluctuations, but also systematically on a macroscopic scale in some types of microsystems. It is only necessary to secure the inviolability of thermodynamic laws on an appreciable scale in the case of usual macroscopic systems encountered in the everyday world.

Quasi-equilibrium systems consisting of plus particles in the presence of minus particles can obviously be attained in the world, provided that we assume that minus particles interact with plus particles only gravitationally and that other forms of interaction (strong, electromagnetic, weak) between plus and minus particles are completely absent. In this case, any system of plus particles can be considered in practice to be adiabatically isolated from minus particles and can therefore be considered to be in a state of thermodynamic equilibrium. Equilibrium can be disrupted only as the result of graviton exchange between systems of plus and minus particles. However, such processes have very low probabilities of occurrence and cannot lead to appreciable effects.

If, in agreement with the assumption made above, only gravitational interactions are possible between plus and minus particles, then minus particles can only be detected in phenomena occurring on a cosmic scale through the gravitational fields produced by large collections of these particles.

It is important to consider another, more reasonable hypothesis which admits all known forms of interaction (strong, electromagnetic, weak, gravitational) between plus and minus particles. Adopting this hypothesis and assuming that minus particles are distributed throughout the universe with a very low concentration (as should be the case in accordance with the results of Section 27), we will not have any appreciable violations of the postulate of thermodynamic equilibrium under terrestrial conditions if we consider only processes of interaction between plus and minus particles. With a concentration of 10^{-6} particles per cm^3, minus particles, interacting with ordinary matter, will not cause larger macroscopic effects under terrestrial conditions than cosmic rays and, therefore, violations of the postulate of thermodynamic equilibrium can only be an effect that is difficult to detect.

The hypothesis of the existence of all forms of interactions between plus and minus particles encounters certain difficulties, however, if we take into account the possibility of processes of multiple creation of groups of plus and minus particles. The usual arguments of quantum field theory do not forbid processes in which two or more plus particles and simultaneously two or more minus particles are created directly out of the vacuum, i.e., so to speak out of nothing. By itself, the process is no more strange than the minus particles themselves. However, in accordance with the usual arguments of quantum field theory, the probability of such a multiple production in vacuo according to perturbation theory increases with the number of particles created.* Therefore, the probability of the creation of objects containing numbers of particles comparable with those in such cosmic objects as stars and galaxies can become so large that such processes will be found to be a very frequent phenomenon.

However, this objection to minus particles possessing all forms of interaction is not as radical as it seems at first sight. Indeed, this objection can be considered as an indication of still another divergence in contemporary quantum field theory, i.e., as one more indication of the inadequacy of the existing theory of elementary particles. It is not surprising that the internal inconsistencies of quantum field theory become so pronounced on the introduction of minus particles, i.e., objects foreign to the existing scheme of quantum theory constructed for positive energies and masses.

––––––––––

* My attention was drawn to this objection to the existence of all forms of interaction between plus and minus particles by T. D. Li in 1964.

The above divergence can be easily eliminated, for example, if we assume that the entropy of the universe of minus particles does not increase, but decreases, i.e., the macroscopic time of minus particles flows in the opposite direction to the time registered by ordinary macroscopic clocks. In this case, the probability of a simultaneous creation of minus particles at a point should be calculated as the probability of their simultaneous collection at one point, i.e., it is a negligibly small quantity. Taking this into account, we find that the probability of the multiple production of a collection of N plus particles and N minus particles cannot be different from the probability of the destruction of such a particle complex in a single act. Consequently, processes of multiple production of plus and minus particles in vacuo cannot lead to any catastrophic violations of equilibrium on a macroscopic scale. It is clear, however, that the hypothesis that time flows in opposite directions in the universe of plus particles and the universe of minus particles must lead to a radical reorganization of quantum field theory.

Hence, we will not consider the hypothesis of the possibility of all known forms of interaction between plus and minus particles to be excluded. Adopting this hypothesis, we obviously admit the possibility of macroscopic violations of classical thermodynamics and recognize the necessity of the development of a new thermodynamics within which classical thermodynamics must only be a special case valid for a narrow class of systems possessing a bounded energy spectrum.

The greatest difficulties, perhaps, are associated with the admission of electromagnetic interactions between plus and minus particles. A minus particle possessing an electric charge would have very strange properties. In passing through ordinary matter consisting of plus particles, a minus particle as the result of Coulomb interactions with electrons and nuclei must continuously lose energy, i.e., it must be accelerated. At the same time, the medium through which the minus particle propagates should become heated. Thus, with a sufficient number of minus particles at our disposal, we could realize an energy source which functions as a thermodynamic machine of the second kind (forbidden by classical thermodynamics), continuously doing work at the expense of a decrease in the energy (cooling) of a heat reservoir consisting of minus particles.

It is obvious that such a machine of the second kind can also be realized with minus particles interacting with ordinary matter through strong and weak interactions. However, the process of energy liberation in the last case will take place very slowly on account of the small cross section for the collision of plus and minus particles. In the case of electromagnetic interactions, energy transfer from minus to plus particles can take place very rapidly because a charged particle initiates many ionizations in the atoms traversed by it. Since such processes have not been discovered up to now in everyday experience, it is reasonable to assume that either charged minus particles do not exist at all, or if they do exist, then there are few of them in our surroundings and the probability of their creation is extremely low. As regards minus particles interacting with plus particles by means of strong and weak forces, we can say that the violation by them of the prohibition of thermodynamic machines of the second kind may only be detectable in phenomena taking place on a cosmic scale and practically undetectable in macroscopic terrestrial experiments.

No matter how remarkable the above consequences of the hypothesis of the existence of minus particles interacting with ordinary matter, physicists have no right to deny their existence only on the basis of *a priori* convictions that macroscopic violations of classical thermodynamics are impossible.

§ 30.   THE DETECTION OF PARTICLES
WITH NEGATIVE MASSES

All known instruments for the detection of particles operate more or less according to the following scheme: A particle enters the instrument from outside and partially or completely loses its energy inside the volume of the instrument; together with this energy, the particle imparts to the instrument a certain amount of negentropy which drives the instrument system away from its initial state of equilibrium (or dynamic quasi-equilibrium); this results in an irreversible cascade-type process which leads to macroscopically detectable effects. This scheme is found in a Wilson cloud chamber, an ionization chamber, a Geiger counter, a Cherenkov counter, nuclear emulsions, etc. It is obvious that particles of negative mass cannot lead to the operation of an instrument working according to this scheme if these particles enter the instrument from outside and are slowed down or absorbed in it. Indeed, a minus particle absorbed in an instrument does not impart energy to it, but extracts energy from it. Consequently, it cannot give rise to the effect that is produced by the absorption of a plus particle, i.e., it cannot impart negentropy to the instrument and thus cannot set it into operation.

The usual type of instrument is obviously suitable for the detection, not of absorption, but of the emission of a particle inside it, since the emitted minus particle will impart to the instrument the same amount of energy as would be imparted to it by an absorbed plus particle. We will also be able to detect the process in which the velocity of the minus particle entering the device will increase. For example, a charged minus particle traversing a Wilson chamber could cause the ionization of condensation centers, but it would lose energy in each single act of ionization, i.e., its velocity would increase. The track left by a minus particle would not be different from that left by a plus particle moving in the opposite direction. Consequently, a charged minus particle entering the Wilson chamber from below, i.e., from inside the earth, would leave exactly the same track as a plus particle that had entered from above, i.e., from outer space. Therefore, charged minus particles produced inside the earth and then accelerated in collisions with plus particles of the earth, on leaving its surface would be registered in a Wilson chamber as charged cosmic-ray particles arriving from space and being slowed down by the material of the earth.

However, as we have established in the preceding section, such electrically charged minus particles either do not exist at all, or for reasons that are as yet unknown, they are produced with a vanishingly small probability, because otherwise the violations of the laws of thermodynamics caused by them would be observable effects.

It is obvious that such processes of autoacceleration of uncharged minus particles cannot be detected by instruments similar to the Wilson chamber. Only the spontaneous emission of a minus particle inside the active region of an instrument such as a Geiger counter could lead to its operation, i.e., could be detected. On the other hand, such instruments are not suitable for the capture and detection of minus particles already in existence.

An instrument intended for the detection of minus particles must be set into operation when the energy in the active region of the instrument is decreased, i.e., as the result of a process equivalent to the emission of a plus particle. Consequently, such an instrument must contain a system which is initially in an intermediate level (not the lowest) and which can go into a lower level only as the result of an absorption of a minus particle, i.e., when negative energy is imparted to it. After the transition to the lower level, an irreversible spontaneous macroscopic process should begin in the system leading to the act of "detection," i.e., the system should have a transition into a third state which is more stable than either of the other two.

Such processes could occur in a system which is in a state with a negative temperature. Indeed, in such a system, the higher energy levels are more stable than the lower ones, inasmuch as the "excitation" of such a system means a transition into a lower energy level and not to a higher one as in the case of ordinary systems with a positive temperature.

Thus, it seems possible in principle to construct instruments capable of detecting the absorption of minus particles.

§ 31. PARTICLES WITH IMAGINARY MASSES AND THE SECOND LAW OF THERMODYNAMICS

It is widely accepted that particles of imaginary proper mass moving with velocities higher than the velocity of light cannot be considered as real objects because the emission and absorption of such particles would involve the transmission of an action from the emitter to the absorber with a velocity higher than that of light. Signalling with a velocity higher than that of light contradicts the Principle of Causality because, according to the theory of relativity, it is always possible to choose a frame of reference in which the time of signal emission (cause) is found to occur later than the absorption of the signal (effect), since these events are connected by a space-like interval. However, as we have established in Section 22, an argument of this type leads to an absolute prohibition of particles of imaginary mass only if the Principle of Causality is considered as an absolute physical law. However, from the point of view of physics, the Principle of Causality is a consequence of the second law of thermodynamics. Consequently, using the same reasoning as that used in connection with particles of negative mass, we can assert that particles of imaginary mass are only forbidden as objects to be used for signalling, although they can appear in processes of the fluctuation type without disrupting the second law of thermodynamics for macroscopic processes of a systematic nature.

It is easy to see that the second law of thermodynamics is not violated in the above process of emission and absorption of a particle of imaginary mass when the probability of emission is equal to the probability of absorption. In this case, it is impossible to distinguish the signal emitter from the receiver and signalling is impossible, because the transfer of the interaction is not of a systematic or directed nature.

The absorption and emission probabilities are obviously equal if space is isotropically filled by particles of imaginary mass, the particles are completely devoid of any charge, and they are absorbed and emitted by ordinary particles of positive mass without a change in the proper mass of the latter.* In this case, as can be seen, for example, from Figs. 14 and 15, the emission and absorption processes are completely symmetric in time and identical, inasmuch as one can be converted into the other by means of a transformation of coordinate systems. Consequently, both the emission and the absorption of a particle of imaginary mass are not accompanied by a change in the total entropy of the system, so that the second law of thermodynamics is not violated.

Particles of imaginary mass do not carry negentropy and cannot be used for signalling, since any signal must carry information, i.e., negentropy.

A somewhat different situation will arise if the particle emitting the particle of imaginary mass changes its mass, charge, spin, etc., during the process, i.e., turns into another particle. In this case, the symmetry in time is violated and the assertion of the

---

*We have shown in Section 26 that it is possible for a particle of positive mass to emit a particle of imaginary mass with only a change in its energy and momentum, and not its rest mass. A similar process involving the emission of a particle of positive mass is not possible.

equality of the emission and absorption probabilities becomes groundless.

Symmetry, however, is restored if we consider, not isolated processes of emission and absorption of particles of imaginary mass, but processes in which a particle of imaginary mass is emitted by particle A and absorbed by particle B in such a manner that particle A changes into particle B and particle B changes into particle A. But this process is physically the same as the well-known process in which momentum, charge, etc., are exchanged between two elementary particles by means of a virtual particle.

Hence, the virtual particles appearing in the quantum theory of elementary particles can be considered as physically real particles with imaginary proper masses exchanged by ordinary elementary particles. The introduction of such particles does not violate the second law of thermodynamics and, consequently, we cannot violate the macroscopic Principle of Causality with their help.

It was shown at the end of Section 26 that particles of imaginary mass can possess either positive or negative energies. Thus, the admission of particles of imaginary mass is not associated with an unavoidable violation of the laws of thermodynamics on a macroscopic scale as was the case with minus particles. The laws of thermodynamics will not be violated microscopically if we forbid particles with negative energies and admit only particles of imaginary mass with positive energies.

§ 32. IS IT POSSIBLE TO DETECT PARTICLES WITH IMAGINARY MASSES?

We have already seen that particles of imaginary mass do not carry negentropy and therefore cannot be used as signals. Thus, it appears that they cannot be detected at all and that they are in this sense unobservable objects.

However, in talking about particles of negative mass, we have already seen that objects exist which cannot be detected by ordinary instruments, but which can be found with the help of measuring devices of a fundamentally new type. We should therefore examine the possibility of the existence of special instruments capable of detecting particles of imaginary mass.

Since the systematic detection of absorption or emission of particles of imaginary mass would lead to the violation of the second law of thermodynamics, we must reject the possibility of the construction of a device capable of detecting a particle of imaginary mass at a given point. This does not mean, of course, that we completely deny the possibility of detecting any effect due to a particle of imaginary mass at a given point, since there is no prohibition on the occurrence of fluctuations in which such particles can collect at one point, the second law of thermodynamics being violated locally, thus leading to the operation of an instrument of the usual type.

Although instruments detecting a particle of imaginary mass at a given point are forbidden, instruments detecting the emission of such a particle at one point and its absorption at another point as a single event are not. Thus, for example, if a particle of imaginary mass carries an electric charge, then the process of its emission by particle A and its absorption by particle B can be detected in nuclear emulsions from the track left by particle A before it emits the particle of imaginary mass and the track of particle B formed after the absorption of the particle of imaginary mass. In other words, it appears possible that we can register the process of charge exchange between charged and neutral particles involving a particle of imaginary mass (i.e., the process which is commonly considered as a process in which a virtual particle is exchanged).

Consequently, particles of imaginary mass can be experimentally detected in principle, although only with the help of special instruments or special experiments in which the processes of emission and absorption of such particles are detected simultaneously.

§33. NEGATIVE MASSES AND NEGATIVE TEMPERATURES

It is well known that systems possessing both a lower and an upper bound to their energy, i.e., an energy spectrum bounded on both sides,

$$E_{\min} < E_n < E_{\max}, \qquad (33.1)$$

can exist in an equilibrium state with a negative absolute temperature.* In fact, the sum over states (29.1) in the case of a bounded energy spectrum converges for both $\Theta > 0$ and $\Theta < 0$. Consequently, the system possesses a canonical equilibrium distribution

$$W(E_n) = Ae^{-E_n/\Theta} \qquad (33.2)$$

with $\Theta > 0$, as well as with $\Theta < 0$, i.e., also with a negative temperature $T = \Theta/k$, since in both cases the normalization condition

$$\sum_k W(E_k) = AZ = 1 \qquad (33.3)$$

is satisfied on account of the convergence of the sum over states.

Quasi-equilibrium states with a negative temperature were discovered experimentally by Purcell and Pound in 1951 [12] in spin systems satisfying condition (33.1).

In the case of a system of plus particles, the energy E_n of the system exceeds the minimum possible energy E_{min} and can be arbitrarily large, i.e.,

$$E_{min} \leqslant E_n < +\infty, \qquad (33.4)$$

since the kinetic energy of the particles can be arbitrarily large.

Θ cannot be negative for such systems, because when $\Theta < 0$ the normalization condition (33.3) cannot be satisfied with a constant A different from zero, on account of the divergence of the sum over states Z. Consequently, if condition (33.4) is satisfied, Θ can only be positive, i.e., $T > 0$. However, in the case of minus particles we obviously have

$$-\infty < E_n \leqslant E_{max} \qquad (33.5)$$

and, consequently, the distribution (33.2) is only meaningful for $\Theta < 0$, i.e., for negative absolute temperatures. Consequently, systems consisting of particles of negative mass can be in a state of

---

*Strictly speaking, this state is a quasi-equilibrium one since no system can be totally isolated from surrounding objects which themselves cannot be in equilibrium at a negative temperature because their energy spectrum is unbounded from above.

thermodynamic equilibrium only at negative absolute temperatures. This question has also been considered in detail by Visin [25].

It can be easily seen that distribution (33.2) cannot be an equilibrium distribution for systems containing particles of positive mass interacting with particles of negative mass. The energy of such systems is unbounded from both above and below and, therefore, the sum of probabilities over all states will diverge for all nonzero values of A. Consequently, such systems cannot be in a state of thermodynamic equilibrium at a definite temperature at all. For them, only quasi-equilibrium states are possible in which the subsystem containing particles of positive mass is in a state of internal equilibrium at a positive temperature, whereas the subsystem consisting of particles of negative mass has a negative temperature. Because of interactions between systems of this type, the temperature of each of them will change and the total system will not be in a state of equilibrium. However, if this interaction is very weak, the disruption of equilibrium will take place very slowly and we can use the concepts of equilibrium thermodynamics for each of the subsystems.

Thus, from the point of view of thermodynamics, systems consisting of particles of negative mass can be considered to possess a negative temperature, provided that the interaction of these systems with ordinary systems of positive mass is sufficiently small.

§ 34. PARTICLES OF NEGATIVE MASS AND COSMIC RAYS

In Section 27 we examined a model of the universe with an average zero rest mass filled approximately uniformly by minus particles at a density of 10^{-30} g/cm^3 and by plus particles collected into stars, planets, interstellar gas, and other cosmic objects. If the minus particles interact with plus particles by means of gravitational, weak, and strong forces,* then they can be considered as a system which is weakly interacting with the system of plus particles, i.e., as a quasi-isolated system. Such a system can be in a state of quasi-equilibrium at a negative temperature. Conse-

---

*We exclude electromagnetic interactions for reasons discussed at the end of Section 29.

quently, in the model under consideration, the universe consists of a system of plus particles nonuniformly distributed throughout space and possessing a positive temperature and a system of minus particles, approximately uniformly distributed and possessing a negative temperature.

It is known from thermodynamics that if we have a heat reservoir in a state with negative temperature, we can realize a thermodynamic machine of the second kind, which can only do work at the expense of the cooling of the single heat reservoir. An example of a machine of the second kind is a laser which produces an intense light flux through the cooling of a quasi-isolated subsystem after the latter has been brought into a state with a negative temperature.*

A thermodynamic machine of the second kind can obviously also be realized with a heat reservoir consisting of minus particles. In contrast to a laser, this machine can do an unlimited amount of work because the heat reservoir consisting of minus particles does not have a finite minimum energy, and can give an unlimited energy output with an unlimited temperature drop. Consequently, in a universe consisting of plus and minus particles we can have a perpetuum mobile of the second kind, i.e., a fundamentally new, practically inexhaustible energy source.

Hence, an essentially new explanation of a number of puzzling astrophysical phenomena, such as the radiation emission from radio galaxies and quasars, becomes possible. These objects emit such vast amounts of energy that it is apparently impossible to explain them in terms of any nuclear or thermonuclear reactions. We now also have a new explanation of the generation of cosmic rays in interstellar and intergalactic space, as well as in the solar atmosphere. Let us investigate the latter in greater detail.

Let $f(E, \mathbf{r}, t)$ be the energy and space distribution of relativistic cosmic particles, ρ_0^+ be the space density of the nonrelativistic plus particles, i.e., the density of ordinary matter (inter-

---

* It should be noted that a laser, being a thermodynamic machine of the second kind, is not a perpetuum mobile of the second kind. The subsystem with a negative temperature feeding the laser has only a limited reserve of energy and in cooling to an arbitrarily low temperature can only produce a limited amount of work.

stellar gas, planets, stars, etc.), ρ^- the space density of minus particles. Colliding with minus particles, a relativistic cosmic-ray particle will on the average increase its energy on account of the increase in the negative kinetic energy of minus particles.* Let us assume that a cosmic-ray particle gains a fraction α of its energy E every second in collisions with minus particles. In other words, we will assume that the average energy of the particle increases according to

$$d\bar{E}/dt = \alpha\bar{E}. \tag{34.1}$$

The coefficient α is obviously proportional to the concentration of minus particles, the velocity of the cosmic-ray particles, and the effective cross section for elastic collisions of plus particles with minus particles σ_{+-}, i.e., it is given by

$$\alpha = n^-c \ \sigma_{+-}, \tag{34.2}$$

where $n^- = \rho^-/m^-$ is the concentration of minus particles, m^- being their mass.

As a result of this process, the average energy of cosmic-ray particles will increase and the equation governing the time variation of the distribution function f will be

$$\frac{\partial f}{\partial t} + \frac{\partial}{\partial E}(\alpha E f) = 0. \tag{34.3}$$

Indeed, according to the last equation for the function f depending only on E and t, we have

$$\frac{d\bar{E}}{dt} = \frac{1}{n^+} \frac{d}{dt} \int_0^\infty Ef\,dE = \frac{1}{n^+} \int_0^\infty E \frac{\partial f}{\partial t}\,dE =$$

$$= -\frac{1}{n^+} \int_0^\infty E \frac{\partial}{\partial E}(\alpha Ef)dE = -\frac{\alpha E^2}{n^+} f\Big|_0^\infty + \frac{\alpha}{n^+} \int_0^\infty Ef\,dE,$$

---

*When a plus particle collides with a plus particle of a stationary medium, its average kinetic energy decreases if the average kinetic energy of the particle in the medium is lower than the kinetic energy of the incident particle, and increases if the kinetic energy of the incident particle is lower than the average kinetic energy of the particles of the medium. On the other hand, in collisions of plus particles with a stationary medium consisting of minus particles, the average kinetic energy of the plus particles always increases.

where

$$n^+ = \int_0^\infty f \, dE$$

is the average number of cosmic-ray particles in a unit volume, i.e., their average concentration. From this, assuming that f is finite at $E = 0$ and that it vanishes sufficiently rapidly as $E \to \infty$, we obtain Eq. (34.1).

The space—energy density f will also change as the result of collisions between cosmic-ray particles and usual matter. If T is the mean free time of a cosmic-ray particle, then the function f must decrease by a factor $1/T$ every second. It is obvious that

$$T = 1 / n_0^+ c\sigma_{+0}, \qquad (34.4)$$

where σ_{+0} is the total effective cross section for the collision of a cosmic-ray particle with particles of nonrelativistic matter; $n_0^+ = \rho_0^+/m^+$, m^+ being the average mass of the plus particles of matter.

The space—energy density f can also change as the result of the usual spatial diffusion of cosmic-ray particles in interstellar magnetic fields.

Taking these two effects into account, as well as the existence of cosmic-ray sources, we have to add three more terms to Eq. (34.3), as a result of which the space—energy equation governing the diffusion of cosmic-ray particles becomes

$$\frac{\partial f}{\partial t} - D\nabla^2 f + \frac{\partial}{\partial E}(\alpha E f) + \frac{f}{T} = Q, \qquad (34.5)$$

where D is the coefficient of diffusion through space, and Q is the density of cosmic-ray sources.

According to [13-16] this equation is a simplified one because it does not take account of changes in particle energies, i.e., it does not contain a term involving the second derivative with respect to E. However, it is adequate for the description of a mechanism for the acceleration of particles similar to Fermi's [17].

The simplest solution of this equation is the time—stationary, spatially uniform solution, i.e., the solution of the equation

$$\frac{\partial}{\partial E}(\alpha E f) + \frac{f}{T} = 0 \qquad (34.6)$$

for energies exceeding the maximum possible energy of the particles generated in the sources. Such a solution is of the form

$$f = AE^{-\gamma}, \tag{34.7}$$

where A is an integration constant and

$$\gamma = 1 + 1/\alpha T = 1 + n_0^+ \sigma_{+0}/n^- \sigma_{+-}. \tag{34.8}$$

It is reasonable to assume that the minus particles have a sufficiently high negative temperature, since they are continually acquiring negative kinetic energies from cosmic-ray particles and cannot accumulate in the form of planets, stars, and other dense cosmic objects maintaining a not too high temperature. Therefore, we can assume that n^- is approximately constant for the whole metagalaxy.* Thus, the coefficient γ must have its minimum value in intergalactic space and increase inside nebulae and close to stars.

The hypothesis of the metagalactic origin of cosmic rays, which easily accounts for their isotropy, corresponds to the assertion that

$$\frac{n_0^+ \sigma_{+0}}{n^- \sigma_{+-}} \simeq 2, \tag{34.9}$$

since we know that $\gamma \simeq 3$ for the primary cosmic-ray component.

Such a relation between the concentrations and cross sections appears to be reasonable if it is considered that the average metagalactic concentrations of plus and minus particles are approximately equal and the effective cross sections σ_{+0} and σ_{+-} are of the same order of magnitude.

Hence, the isotropy and the energy spectrum of the primary component can be easily explained if it is assumed that cosmic

---

*The concentration of minus particles may increase inside stars, since, according to the Boltzmann distribution, at negative temperatures we have

$$n^- \sim \exp(-m\varphi/kT) = \exp(-|m|\varphi/k|T|),$$

where φ is the Newtonian gravitational potential. However, this increase should be negligibly small at high $|T|$.

rays are formed in intergalactic space and reasonable values are adopted for the concentration of minus particles and the cross section for the interaction between minus and plus particles [18].

The concentration of plus particles n_0^+ increases in the vicinity of stars and, if it is assumed that the concentration of minus particles n^- remains practically constant, then, according to (34.8), the index γ should increase. Thus, cosmic rays should also be generated in the vicinity of the sun, although their spectrum should be steeper than that of cosmic rays generated by the metagalaxy as a whole. The intensity of energy liberation, i.e., the amount of energy acquired by cosmic-ray particles per unit volume per unit time is obviously proportional to the concentration n^+ of cosmic-ray particles and the energy increment per unit time, i.e., according to (34.1) we have

$$dW / dt = \alpha \overline{E} n^+, \qquad (34.10)$$

where W is the energy density of cosmic-ray particles.

Thus, with a coefficient α approximately constant for all regions of space, the energy liberation increases proportionally to the concentration of cosmic-ray particles. Consequently, if some region of space, for example, the vicinity of a supernova, contains an accumulation of cosmic-ray particles, then energy liberation in this region will also be increased.

In order to account for the radio emission of radio galaxies, it is sufficient to assume that relativistic particles have accumulated in them as the result of some process with a concentration several orders of magnitude higher than the concentration of gas in intergalactic space, so that at the given time the intensity of the total radio radiation emitted by these particles in the magnetic fields present in the radio galaxies is equal to the observed value. This energy liberation can be maintained at a constant level by collisions between the relativistic particles and minus particles as can be seen from formula (34.10). Indeed, with $n^- \sim 10^{-6}$ cm^{-3}, $\sigma_{+-} \sim 10^{-26}$ cm^2, $c = 3 \cdot 10^{10}$ cm/sec, we obtain $\alpha \sim 3 \cdot 10^{-22}$ sec^{-1}. Assuming next that $\overline{E} \sim 10^{10}$ eV $\simeq 1.6 \cdot 10^{-2}$ ergs, $n^+ \sim 10^{-3}$ cm^{-3}, we find for a radiogalaxy with a volume of 10^{68} cm^3 that the total power radiated is $\sim 5 \cdot 10^{41}$ ergs/sec which agrees with the observed value. If the concentration of relativistic particles is increased within reasonable limits, the power released can be considerably increased. Thus, the hypothesis under consideration also allows us to explain in a reasonable manner the colossal energy liberation in quasars without recourse to the questionable hypothesis of gravitational collapse.

A similar hypothesis explaining the energy liberation in quasars has been proposed independently by Banesh Hoffmann [20, 21]. He suggested that an intense process of creation of particles of negative mass and consequently a positive yield of energy takes place in quasars.

LITERATURE CITED

1. A. Einstein, Ann. Phys., 17:891 (1905).
2. H. Poincaré, Rend. Circolo Mat. Palermo, 21:129 (1906).
3. H. Minkowski, Phys. Z. S., 10:104 (1909).
4. L. I. Mandel'shtam, Collected Works, Vol. V, Izd. Akad. Nauk SSSR (1950).
5. H. A. Lorentz, Proc. Acad. Sci. Amsterdam, 6:809 (1904).
6. R. V. Serebryanyi, Zh. Eksperim. i Teor. Fiz., 20:1130 (1950).
7. Ya. P. Terletskii, Dynamical and Statistical Laws of Nature, Izd. MGU (1950).
8. Ya. P. Terletskii, Statistical Physics, Izd. "Vysshaya shkola" (1966).
9. Ya. P. Terletskii, Dokl. Akad. Nauk SSSR, 133:329 (1960).
10. Ya. P. Terletskii, J. Phys. et Radium, 21:681 (1960); 23:910 (1962).
11. P. A. M. Dirac, Proc. Roy. Soc., 117:610 (1928); 118:341 (1928).
12. E. M. Purcell and R. V. Pound, Phys. Rev., 81:279 (1951).
13. Ya. P. Terletskii and A. A. Logunov, Zh. Eksperim. i Teor. Fiz., 21:567 (1951).
14. A. A. Logunov and Ya. P. Terletskii, Izv. Akad. Nauk SSSR, ser. fiz., 17:119 (1953).
15. P. Morrison, S. Olbert, and B. Rossi, Phys. Rev., 94:440 (1954).
16. Ya. P. Terletskii, Dokl. Akad. Nauk SSSR, 101:59 (1955).
17. E. Fermi, Phys. Rev., 75:1169 (1949).
18. Ya. P. Terletskii, Ann. Inst. Henri Poincaré, 1:431 (1964); Quasi-Stellar Sources and Gravitational Collapse, Chicago (1965), p. 466.
19. O. Costa de Beauregard, Proceedings of the 1964 International Congress for Logic Methodology and Philosophy of Sciences, Jerusalem, Aug. 26-Sept. 2, 1964 (North Holland Publ. Co., Amsterdam).

20. Banesh Hoffmann, Science, April 1965, p. 75.
21. Banesh Hoffmann, "Negative mass and the quasars," in: Perspectives in Relativity and Differential Geometry, Indiana University Press (1966).
22. S. Tanaka, Progr. Theoret. Phys. (Kyoto), 24:171 (1960).
23. G. Feinberg, Phys. Rev., 159:1089 (1967).
24. O. M. P. Bilaniuk, V. K. Deshpande, and E. C. G. Sudarshan, Am. J. Phys., 30:718 (1962).
25. V. Visin, Physics Letters, 2:32 (1962); 3:174 (1963); 13:217 (1964).

INDEX

119